T0325161

Solar Powered Wastewater Recycling

The United Nations predicts that by the year 2025, two-thirds of the world's population will face water scarcity. Further, the planet would have well over eight billion people, the majority of whom would live in developing countries, where more than 80% of those are already experiencing water scarcity. Therefore, there is an urgent need for wastewater recycling to help solve issues of scarcity and to facilitate better management of generated wastewater. Water recycling includes reuse and treatment of municipal wastewater, which could be a sustainable approach for environmental sustainability and could also help to offset the increasing water demands for irrigation and industrial and other needs. Currently, water and wastewater treatment facilities consume large amounts of energy that are mainly generated through the use of fossil fuels. *Solar Powered Wastewater Recycling* examines how solar power can be implemented as an integrated approach whereby all the energy needs of the water and wastewater sector could be supplemented by renewable technologies, and in which a synergy can be developed between water and energy.

Solar Powered Wastewater Recycling

Arun Kumar
and Sandhya Prajapati

CRC Press
Taylor & Francis Group
Boca Raton London

CRC Press is an imprint of the
Taylor & Francis Group, an **informa** business

Designed cover image: Shutterstock

First edition published 2024
by CRC Press
6000 Broken Sound Parkway NW, Suite 300, Boca Raton, FL 33487–2742

and by CRC Press
4 Park Square, Milton Park, Abingdon, Oxon, OX14 4RN

CRC Press is an imprint of Taylor & Francis Group, LLC

Library of Congress Cataloging-in-Publication Data

Names: Kumar, Arun (Microalgist), author. | Prajapati, Sandhya, author.
Title: Solar powered wastewater recycling/Arun Kumar and Sandhya Prajapati.
Description: First edition. | Boca Raton: CRC Press, [2023] | Includes bibliographical references and index.
Identifiers: LCCN 2023001162 (print) | LCCN 2023001163 (ebook) | ISBN 9781032526508 (hbk) | ISBN 9781032526515 (pbk) | ISBN 9781003407690 (ebk)
Subjects: LCSH: Water treatment plants-Power supply. | Solar power plants. | Water-Purification-Energy conservation. | Water reuse.
Classification: LCC TD434 K86 2023 (print) | LCC TD434 (ebook) | DDC 628.1/620286-dc23/eng/20230131
LC record available at https://lccn.loc.gov/2023001162
LC ebook record available at https://lccn.loc.gov/2023001163

ISBN: 978-1-032-52650-8 (hbk)
ISBN: 978-1-032-52651-5 (pbk)
ISBN: 978-1-003-40769-0 (ebk)

DOI: 10.1201/9781003407690

Typeset in Times
by Apex CoVantage, LLC

Contents

Abbreviations

AD	Accelerated Depreciation
ADB	Asian Development Bank
AfD	Agence Française de Développement
AMRUT	Atal Mission for Rehabilitation and Urban Transformation
AOPs	Advanced oxidation processes
AP	Anaerobic pond
ASP	Activated sludge process
BCM	Billion cubic meters
BEE	Bureau of Energy Efficiency
BIOFOR	Biological filtration and oxygenated reactor
BIPV	Building-integrated photovoltaic
BOD	Biological oxygen demand
CCGT	Combine cycle gas turbine
CEA	Centre Electricity Authority
CEC	California Energy Commission
CFU	Colony Forming Unit
CGWB	Central Ground Water Board
COD	Chemical oxygen demand
CETPs	Common effluent treatment plants
CPCB	Central Pollution Control Board
CPHEEO	Centre Public Health Environment and Engineering Organization
CPSU	Central public sector undertaking scheme
CSF	Constructed soil filter
CSP	Concentrating solar thermal power
CW	Constructed wetland
DC	Direct current
EA	Extended aeration
EE	Energy efficiency
EPA	Environmental Protection Agency
FAB	Fluidized aerobic bioreactor
FAO	Food and Agriculture Organization
FBAS	Fixed bed bio-film activated sludge
FP	Facultative pond
FYP	Five-year plan
GDP	Gross domestic product
GHG	Greenhouse gases
GRIHA	Green rating for integrated habitat assessment
GW	Gigawatt
HVAC	Heating, ventilation and air conditioning
IGBC	Indian Green Building Council
INDC	Intended Nationally Determined Contributions
INR	Indian rupees

ISO	International Standard Organization
JICA	Japan International Cooperation Agency
KfW	Kreditanstalt für Wiederaufbau
LED	Light-emitting diode
LPCD	Liters per capita per day
MBBR	Moving bed bioreactor
MBR	Membrane bioreactor
MGI	McKinsey Global Institute
MLD	Million liters per day
MoUD	Ministry of Urban Development
MoWR	Ministry of Water Resources
MWh	Million gigawatts per hour
NPK	Nitrogen phosphorous potassium
NRW	Non-revenue water
O&M	Operations and maintenance
PV	Photovoltaics
RAS	Return activated sludge
RCC	Reinforced cement concrete
SAFF	Submerged aeration fixed film
SBM	Swachh Bharat Mission
SBR	Sequencing batch reactor
SBT	Soil bio-technology
STPs	Sewage treatment plants
TOC	Total organic content
UASB	Upflow anaerobic sludge blanket
ULB	Urban local body
UN	United Nations
UNICEF	United Nations International Children's Emergency Fund
UN	United Nations
USA	United States of America
USD	U.S. dollar
UV	Ultraviolet
VGF	Viability gap funding
WAS	Waste activated sludge
WB	World Bank
WII	Winrock International
WSP	Waste stabilization ponds

About the Authors

Dr. Arun Kumar is currently working as Assitant Professor in Roorkee College of Pharmacy (PG College of Allied Science). Dr. Kumar has published various research and review papers in reputed journals like *Frontiers in Microbiology*, *Frontiers in Biosciences* and *Environment and Sustainability Indicators* in the area of microalgal bioremediation, biofuels and bioproducts production from microalgae. He also published an international book *Microalgae in Wastewater Remediation* from CRC Press, Taylor & Francis Group. He obtained his doctoral degree in environmental microbiology from Babasaheb Bhimrao Ambedkar University, Lucknow, India. In his doctoral work, Dr. Kumar successfully metabolized a widely used and hazardous insecticide Chlorpyrifos by cyanobacterial strain *Oscillatoria* sp. CYA8 CPF isolated from paddy fields.

Dr. Sandhya Prajapati is currently working as Assistant Professor in the Department of Electrical Engineering at Dev Bhoomi Uttarakhand University. Dr. Prajapati has published various research papers "Energy Sources part A: Recovery, Utilization and Environmental Effects," *International Journal of Renewable Energy Technology* (Inderscience) and *International Journal of Sustainable Engineering* in load management, capacity credit and electric vehicle charging strategies. He obtained his doctoral degree in electrical engineering from Indian Institute of Technology, Roorkee, India. In his doctoral work, Dr. Prajapati successfully studied utilization of solar energy in various fields, especially in greywater recycling.

1 Introduction

INTRODUCTION

Due to ever-increasing water demand and consequent pollution of water resources, water recycling and reuse is becoming the uttermost need of current world, dealing with both the issues of: (a) water scarcity; and (b) water pollution. It is reported that water use increased six-fold during the 20th century, while the water availability is constantly decreasing at global level. It is predicted that two-thirds of the world's population will face water scarcity by 2025, while the annual water supply will be of less than 1700 m^3 per capita. Due to this, about 1.8 billion people will forced to undergo absolute water scarcity conditions with an annual water supply of less than 500 m^3 per capita. Further, the planet earth would have eight billion people in 2030, of which most of the population will live in developing countries, where 82% of the world's population are already experiencing water scarcity (World Bank, 2010). It will have increased the demand for water in these countries that is mainly fulfilled by the groundwater, which could reduce the water availability for irrigation purposes and have a negative impact on food production (FAO, 2007).

India accounts for 16% of the world's population but has merely 4% of the world's fresh water resources. Due to changing weather patterns and recurring droughts, groundwater is becoming the only option. It is reported that the groundwater in as many as 256 of 700 districts reached at critical or over exploited level (CGWB, 2017). Further, there is estimation that more than 50% of the country's population will resided in cities and towns by 2050; this leads to the two everlasting situations: i.e., water scarcity and sewage overload. There is lack of proper sewerage network and wastewater treatment facilities in the majority of the towns and cities. Further, there is unplanned expansion of towns and cities beyond their municipalities that still remained under rural administrations which lack the infrastructure and finances required for the management of municipal wastewater. It is further worsened in smaller towns, where municipal wastewater is either discharged into rivers or lakes or in open fields.

STATUS OF WASTEWATER GENERATION IN INDIA

There is estimation of about 38,254 MLD (million liters per day) of wastewater produced from the urban areas including Class-I cities (having population greater than 100,000) and Class-II towns (having population in range of 50,000–100,000), which make up the more than 70% of the total urban population of India. But the treatment capacity of about 11,787 MLD of municipal wastewater is generated so far, that is only 31% of total wastewater produced in these Class-I cities and Class-II towns (CPCB, 2009a). It will further have projected to 120,000 MLD of municipal wastewater produced by 2051, while rural India will not produce so much wastewater but

DOI: 10.1201/9781003407690-1

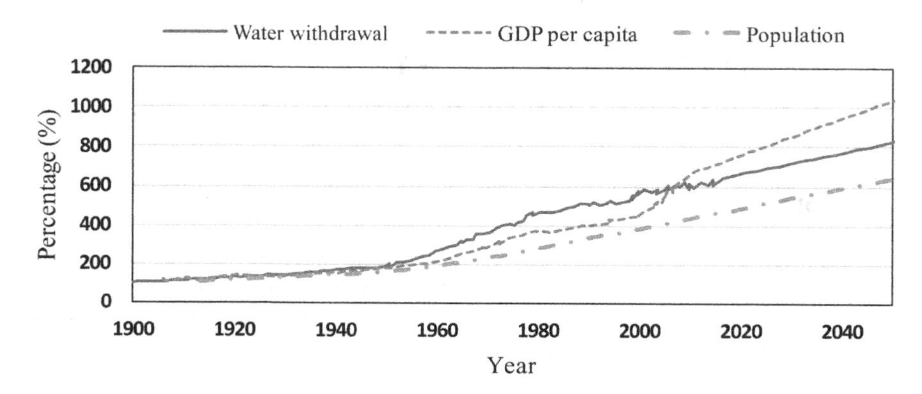

FIGURE 1.1 World population and water shortage (Boretti and Rosa, 2019)

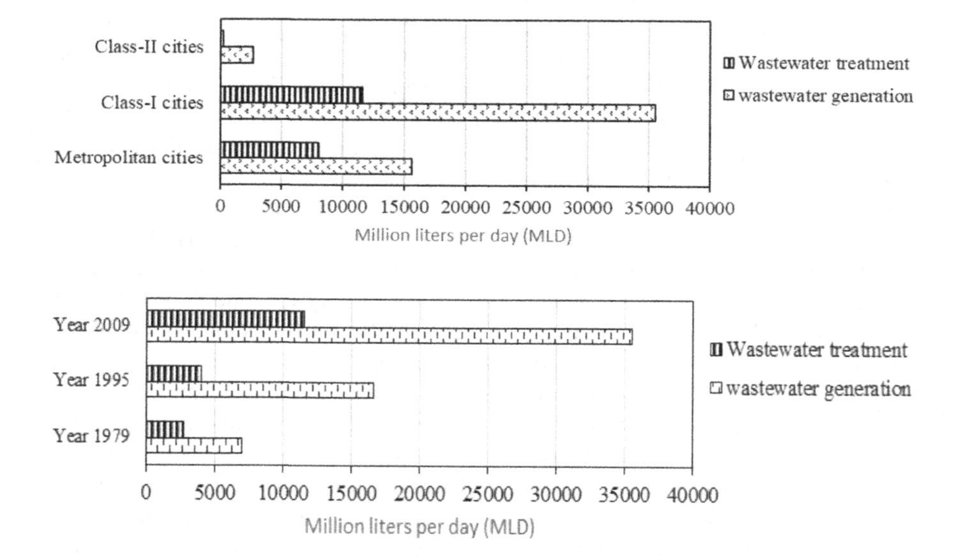

FIGURE 1.2 Wastewater generation trends in India (CPCB, 2013)

not less than 50,000 MLD due to upcoming planned water supply for communities in rural areas. This volume of municipal wastewater would be not possible for the current infrastructure and management plans.

CPCB (2009a) underlined the reality of the existing treatment plants and observed that they are not working at their full capacity, only operating at about 72% of operating capacity. It is estimated that more than 75% of the municipal wastewater that generated in Class-I and Class-II urban towns and cities is consequently released untreated into water bodies or fields, which leads to the enormous environmental pollution. This environmental pollution not only poses serious public health issues but also is responsible for the contamination of 80% of the surface water available in

India. Due to insufficient sanitation and unhygienic practices, there could be social impact and massive economic loss to the nation, which was roughly about INR 2.4 trillion or USD 53.8 billion (equivalent to about 6.4% of India's GDP) in the year of 2006 (WSP, 2011). These economic losses and social impacts primarily burden the poorer sections of society due to their densely populated areas, poor hygiene practices, interrupted water supply and less access to the improved sanitation.

NEED AND SCOPE OF WATER RECYCLING

India is currently growing at population growth of 1.7% per annum with urbanization rate of 3% per decade, and it is predicted that urban population of 377 million in 2011 will reach 590 million in 2030 (MGI, 2010). Considering this, there would be nearly 60,000 MLD of municipal wastewater produced from urban centers by 2030; from this, if 80% of the produced municipal wastewater could undergo treatment, there would be roughly a 400% increase in the availability of treated wastewater; that would be around 17 BCM treated wastewater recycled per year. This 17 BCM of recycled wastewater could meet 75% of the projected water demand of industrial sector and 25% total proposed drinking water demand in 2025 (MoWR, 2006).

The increasing trend of urbanization in India poses a significant challenge to developing the urban infrastructure and services such as water supply, sanitation, disposal of solid waste, adequate drainage systems and collection and treatment of wastewater. Insufficient sanitation infrastructure and poor hygiene practices are responsible for enormous economic and social losses. It is estimated that an annual loss of nearly INR 2.4 trillion (US 53.8 billion) occured due to inadequate sanitation in the year 2006, which was roughly about 6.4% of India's GDP in that year (WSP, 2011). These economic impacts and losses are borne excessively by poor people who live in densely populated areas which have limited water supply and improved sanitation.

Water recycling includes reuse and treatment of municipal wastewater (including both greywater and blackwater), and could be a sustainable approach for the environmental protection – and it also fulfills the increasing water demands for irrigation and the industrial sector. Most of the water utilized for non-potable needs – whether in the domestic sector (in toilet flushing, bathing, washing, etc.), for agriculture or in industry – represents a tremendous opportunity to recycle the water by giving it various levels of treatment.

BENEFITS OF WATER RECYCLING

AN ADDITIONAL SOURCE OF WATER

Recycled water could be an additional source of water other than the groundwater, surface water and rain-fed water, which could be a potential option for irrigation and industrial purposes. It is evident that drinking water supplied from lakes or reservoirs or through canals, which is quite distant from the major cities and towns, resulted in the increase in the cost of water supply. This recycled water could reduce total water demand by diverting the non-potable water demand to recycled water.

Due to steep and rising water tariffs, a significant amount of money is used up by industrial enterprises. Here recycled water could be a reliable source, and it could be more affordable water supply than the freshwater supply in most of the scenarios.

Agriculture mainly depends upon rainwater, which is affected by the changing weather patterns, for irrigation, resulting in the huge pressure on groundwater resources. Thus, recycled water not only provides a greatly needed alternative for irrigational purposes, but also helps in groundwater recharging. In India, nearly 38,254 MLD of wastewater produced from the urban areas could be able to provide about 14 BCM of recycled water, which would be sufficient to irrigate an area about of 1–3 million hectares (ha) (CPCB, 2009a). The irrigation potential of this recycled water could be nearly three times more than the surface water-based irrigation potential and also much as 44% of the major- and medium- irrigation potential (taken at 2 million ha) developed during 10th five-year plan (FYP).

Source of revenue for municipalities

Municipalities are the primarily engaging in the establishment and management of sewerage and treatment plants (STPs) in the cities and towns that require so much money for the efficient operations and maintenance of these facilities. If these utilities ensured the proper and maximum collection of municipal wastewater and STPs are adequate and in well functioned states, then these utilities are in a position to sell the recycled water to industries according to their needs and availability of the water from other sources. Municipalities can decide water tariffs for recycled water according to treatment level provided and the quality required by industry.

Unlike concessional water tariffs for urban use and irrigational purposes, municipalities can recover the actual cost spent for the treatment and supply of the recycled water – which includes maximum share of O&M (operations and maintenance) costs. WSP (2014) conducted a study in Chennai, in which treated wastewater sold at INR 8–11 KL^{-1} (USD 0.13–0.1823) to the industries, and it is sufficient to cover the O&M costs of the treatment plants. Further, municipalities need to encourage industry for the use and sale of recycled water to industrial customers; if required changes can be made in state and local regulations to make this practice mandatory. This recycled water from Class-I cities and Class-II towns would be able to meet about a quarter of the current industrial water demand (i.e., 17 BCM including the water demand for energy production) by 2030.

Nutrient recycling

Being an additional water resource, municipal wastewater also poses significant amounts of nitrogen, phosphorus and potassium (NPK) that can help reduce the need for synthetic fertilizers up to 40% in India (Minhas, 2002; Silva and Scott, 2002; Kaur et al., 2012). It is estimated that municipal wastewater has daily nutrient potential of 0.054–0.073 tons MLD^{-1} wastewater (adapted from Minhas, 2002; Silva and Scott, 2002; CPCB, 2009a; WII, 2006); thus, an estimated nutrient load of about 2,500 tons day^{-1} could be procured from the total wastewater produced in Class-I and

Class-II cities of India. According to CPCB (2009a) estimation, there could be nutrients of value of INR 8,000 (USD 1653) ton^{-1}; considering this at present, the total potential of wastewater produced in urban areas of the country could be worth of about INR 19.5 million or USD 0.4 million (as per INR 500 or USD 10.334 MLD^{-1}). From the analysis of studies by WII (2006), Londhe et al. (2004) and Amerasinghe et al. (2013), it is also suggested that using recycled water for irrigation could help in increasing a 30% increase in annual farm income to farmers.

Reduction in energy required for groundwater pumping

Besides with rainwater, agriculture fulfills its demand by fetching or pumping groundwater for irrigation that requires much energy, which leads to the increase in the cost of agricultural yield. Using recycled water, the amount of groundwater needed for irrigation could be reduced, and also there could be a decrease in the demand for groundwater pumping–related energy, which ultimately helps achieve the goal of reduction in the harmful greenhouse gas (GHG) emissions that are released during the production of electricity. In place of groundwater, the adoption of recycled water for irrigation could be helpful in conserving the electricity up to 1.75 million MWh, equivalent to about 1.5 million tons of CO_2 (tCO_2) GHG emissions which are released during the electricity production.

WASTEWATER TREATMENT

Conventional wastewater treatment involves a series of processes which are primarily categorized in to preliminary, primary, secondary, and tertiary and disinfection processes. These processes produce large quantities of various types of solids and sludge that undergoes further handling, management and treatments (Kumar and Singh, 2021).

Preliminary treatment primarily includes the use of mechanical or manually-cleaned bar screens to remove large solids, known as *screening*, followed by removal of the grit and scum; these activities are facilitated in *grit removal tanks*.

After this, it is necessary to make wastewater the homogenous and aerated, needing *primary* treatment whereby wastewater is properly mixed with provision of air flow. Then mixed and aerated wastewater are conveyed to a *primary sedimentation tank* where *scum* floats to the surface, scraped through, and *sludge* settles down to the bottom into what is known as *primary sludge*. The sludge is pumped out and sent to sludge handling facility, and the treated wastewater is conveyed to secondary treatment.

The primary treated wastewater still has the organic matter, responsible for the biological oxygen demand (BOD), so the wastewater undergoes *secondary* treatment which involves biological treatment. In this process, micro-organisms – mainly bacteria – degrade the organic matter in to sludge and scum (suspended solids), followed by secondary sedimentation tank where sludge and scum settles and is collected as *secondary sludge*.

The primary and secondary treated wastewater still contains color, dissolved solids and toxic compounds that are removed through *tertiary treatment*. Tertiary treatment involves a number of techniques, mainly filtration. Finally, treated wastewater undergoes processes like chlorination and ozonation that kills the microbes, especially pathogenic microbes presented in the wastewater.

Sludge processing involves facilities for the handling, management and processing of the sludge that takes place and produces compost, ashes or other end products. The characteristics of the end products primarily depend up on sludge type (primary, secondary or mixed), their treatment and the processes used for the treatment. Depending on the wastewater treatment process, sludge could be classified in to: (a) primary sludge produced during primary wastewater settling; (b) secondary sludge or activated sludge produced during secondary biological treatment; (c) mixed sludge produced from the mixing of primary and secondary sludge; and (d) tertiary sludge produced during tertiary or advanced wastewater treatment.

THE ENERGY PROBLEM IN WATER RECYCLING – AND THE SOLUTION

Water and wastewater treatment facilities include a network of pump stations, distribution lines and storage facilities to transfer the water flow to various unit operations and treatment processes. Further, they have included various treatment processes such as sedimentation, coagulation, flocculation, filtration and disinfection processes that often vary from plant to plant. In all cases, water and wastewater treatment facilities consume huge volumes of energy that are mainly fulfilled by the grid electricity, a significant amount of which is generated through use of fossil fuels. It is estimated that water and wastewater treatment facilities in the United States are responsible for the consumption of about up to 75 billion kWh of energy for the treatment operations and also are responsible for emission of 45 million tons of GHGs in the air (EPA, 2009).

There are two approaches for reducing the energy consumption related to water and wastewater management systems: (a) energy efficiency (EE); and (b) use of clean energy like solar energy (Barry, 2007). For energy efficiency, pumping and aeration systems could be optimized, leading to reduction in energy consumption related to water and wastewater treatment; whereas the utilization of solar power – along with energy efficiency – provides a clean alternative that not only fulfills energy needs but also helps reduce the carbon emissions.

SOLAR POWERED WASTEWATER RECYCLING (SPWR)

Solar powered wastewater recycling is an integrated approach whereby all the energy needs of the water and wastewater sector are supplemented by solar energy (Prajapati and Fernandez, 2019a, 2021b). Through this approach, a synergy can be developed between water and energy which not only reduces the water treatment–related electricity consumption (from the grid responsible for carbon emissions) and the water requirement of the non-potable activities can be fed with the recycled water (Prajapati and Fernandez, 2019a, 2021b). Most of the water supply and wastewater treatment activities follow in the day period when solar energy is available, and can be done without utilization of additional electricity or battery backup.

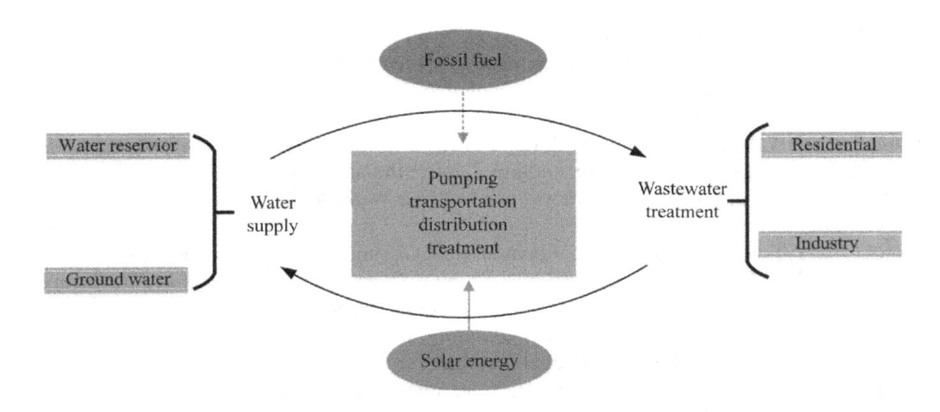

FIGURE 1.3 Concept and objectives of SPWR

There are two options: solar thermal and solar electricity that used for harnessing the solar energy (Prajapati and Fernandez, 2019b, 2019c, 2020a, 2020b, 2021a). Solar thermal electricity is suitable for the heating and cooling purposes and could be helpful in desalinization of salt water; whereas solar electricity involves the use of PV (photovoltaics) that capture the solar energy and convert it in to electricity that is not only useful for heating and cooling but is also able to run the pumps and equipment that are integral part of the water and wastewater treatment activities. These PVs can be mounted on the rooftop of houses, offices and buildings and be further supplemented with grid or battery backup for better utilization of solar power.

REFERENCES

Amerasinghe, P., R. M. Bhardwaj, C. Scott, K. Jella, and F. Marshall. 2013. "Urban Wastewater and Agricultural Reuse Challenges in India." In *IWMI Research Report*. Colombo: IWMI 147: 36p.

Barry, J. A. 2007. *WATERGY: Energy and Water Efficiency in Municipal Water Supply and Wastewater Treatment. Cost-Effective Savings of Water and Energy. The Alliance to Save Energy*. New York: U.S.AID.

BEE. 2011. "Mitigation Initiatives Through Agriculture Demand Side Management." Presentation made in 2011 by Sarabjit Singh Saini, Bureau of Energy Efficiency, Ministry of Power, Government of India, August.

Boretti, A., and L. Rosa. 2019. "Reassessing the Projections of the World Water Development Report." *npj Clean Water* 2, no. 1: 1–6.

CGWB. 2017. *Dynamic Ground Water Resources of India (As on March 2013)*. Ministry of Water Resources, Govt. of India. http://cgwb.gov.in/Documents/Dynamic%20GWRE-2013.pdf

CPCB. 2009a. *Status of Water Supply, Wastewater Generation and Treatment in Class-I Cities and Class-II Towns of India*. New Delhi: CPCB, Ministry of Environment and Forests, Government of India.

CPCB. 2009b. *Comprehensive Environmental Assessment of Industrial Clusters*. New Delhi: CPCB, Ministry of Environment and Forests. Government of India.

CPCB. 2013. *Performance Evaluation of Sewage Treatment Plants Under NRCD*. New Delhi: Government of India.

EPA. 2009. *National Primary and Secondary Drinking Water Regulations*. EPA 816-F09-004. Washington: United States Environmental Protection Agency.

FAO. 2007. *Coping with Water Scarcity*. Challenge of the Twenty-First Century, UN-Water. http://www.worldwaterday07.org.

Kaur, R., S. P. Wani, A. K. Singh, and K. Lal. 2012. *Wastewater Production, Treatment and Use in India*. New Delhi: Water Technology Centre, Indian Agricultural Research Institute.

Kumar, A., and J. S. Singh. 2021. *Microalgae in Waste Water Remediation*. CRC Press, Boca Raton.

Londhe, A., J. Talati, L. K. Singh, M. Vilayasseril, S. Dhaunta, B. Rawlley et al. 2004. "Urban-Hinterland Water." In *Transactions: A Scoping Study If Six Class I Indian Cities*. Paper Presented at IWMI-Tata Annual Partners Meeting, Anand, India.

MGI. 2010. *India's Urban Awakening: Building Inclusive Cities, Sustaining Economic Growth*. London: McKinsey Global Institute.

Minhas, P. S. 2002. *Use of Sewage in Agriculture: Some Experience. Presentation at Workshop on Wastewater Use in Irrigated Agriculture: Confronting Livelihoods and Environmental Realities*. Hyderabad: IWM.

MoWR. 2006. *Report of the Working Group on Water Resources for the XI Five Year Plan (2007–2012)*. New Delhi: MoWR (Ministry of Water Resources).

Prajapati, S., and E. Fernandez. 2019a. "Residential Load Management with Gray Water Recycling to Maximize Rooftop Solar PV Usage." *Energy Sources, Part A*: 1–13. https://doi.org/10.1080/15567036.2019.1687618.

Prajapati, S., and E. Fernandez. 2019b. "Rooftop Solar PV System for Commercial Office Buildings for EV Charging Load." *International Conference on Smart Instrumentation, Measurement and Application (ICSIMA) IEEE*: 1–5. https://doi.org/10.1109/ICSIMA47653.2019.9057323.

Prajapati, S., and E. Fernandez. 2019c. "Estimation of Maximum Energy Storage Levels for an Isolated Rural Microgrid Using Monte Carlo." *International Conference on Electrical, Electronics and Computer Engineering (UPCON) IEEE:* 1–4. https://doi.org/10.1109/UPCON47278.2019.8980040.

Prajapati, S., and E. Fernandez. 2020a. "Reducing Conventional Generation Dependency with Solar Pv Power: A Case of Delhi (India)." *Energy Sources Part A*: 1–12.

Prajapati, S., and E. Fernandez. 2020b. "Solar PV Parking Lots to Maximize Charge Operator Profit for EV Charging with Minimum Grid Power Purchase." *Energy Sources, Part A*: 1–11. https://doi.org/10.1080/15567036.2020.1851325.

Prajapati, S., and E. Fernandez. 2021a. "Capacity Credit Estimation for Solar PV Installations in Conventional Generation: Impacts with and Without Battery Storage." *Energy Sources, Part A* 43, no. 22: 2947–59. https://doi.org/10.1080/15567036.2019.1676326.

Prajapati, S., and E. Fernandez. 2021b. "Standalone Solar PV System for Grey Water Recycling Along with Electric Load for Domestic Application." *International Journal of Sustainable Engineering* 14, no. 5: 933–40. https://doi.org/10.1080/19397038.2020.1739168.

Silva, P., and C. Scott. 2002. *What Are the Wastewater Treatment Plant Effect on Wastewater Irrigation Benefits?* Presentation at Workshop on Wastewater Use in Irrigated Agriculture: Confronting Livelihoods and Environmental Realities. Hyderabad: IWMI.

WII. 2006. *Urban Wastewater: Livelihoods, Health and Environmental Impacts in India*. New Delhi: Winrock International.

World Bank. 2010. *Sustaining Water for All in a Changing Climate*. Washington, DC: World Bank: 57125.

WSP. 2011. *Economic Impact of Inadequate Sanitation in India*. New Delhi: Water and Sanitation Program.

WSP. 2014. *Cost of Wastewater Treatment Recovery in Sewerage and Sewage Treatment*. New Delhi: Water and Sanitation Program.

2 Wastewater treatment
On-site systems

For an efficient and well-defined sanitation system, some basic functions to address include: proper collection of wastewater or waste; safe transport of that wastewater or waste, either by drains or sewers; and then efficient treatment of waste or wastewater, before reusing it or discharging to the environment (Carr and Strauss, 2001). The sewerage network in India starts with the collection of sewage from source of generation, conveying it through a sewer network, followed by the various treatment technologies and finally discharging to natural water bodies.

Under the Swachh Bharat Mission (SBM) initiative, India has shown remarkable improvement in sanitation coverage in terms of access to toilets; in rural areas, nearly 93% of households have the toilets, while 96.2% of the households in the urban areas have access to toilets in 2019 (Ministry of Jal Shakti, Government of India, 2020). Through universal access to toilets, a few issues can be resolved like collection of waste or wastewater, temporary storage through on-site systems, and partial treatment; but still there is a big issue of reaching the centralized sewer network to increase safe collection and treatment. It is reported that only 39% of households are connected with a sewer network, while 48.9% have on-site systems like septic tanks.

WASTEWATER COLLECTION: OVERVIEW AND TYPES

Wastewater collection of particular types – i.e., greywater, combined greywater and septic tank effluent or sewage and stormwater – is an intervention for safe and efficient wastewater treatment (Figure 2.1). Wastewater collection systems fall into the following three categories.

1. Covered surface drains
2. Small-bore sewers
3. Conventional sewers

COVERED SURFACE DRAINS

Covered surface drains are more adapted in the rural areas of India, where per capita water supply is found to be 40–70 LPCD, leading to less wastewater generation (Mehta et al., 2013). This low amount of wastewater is difficult and unsuitable for effective operation of conventional sewer systems. There proved to be the two following cost-effective and interim approaches for collection and treatment of following water/wastewater.

DOI: 10.1201/9781003407690-2

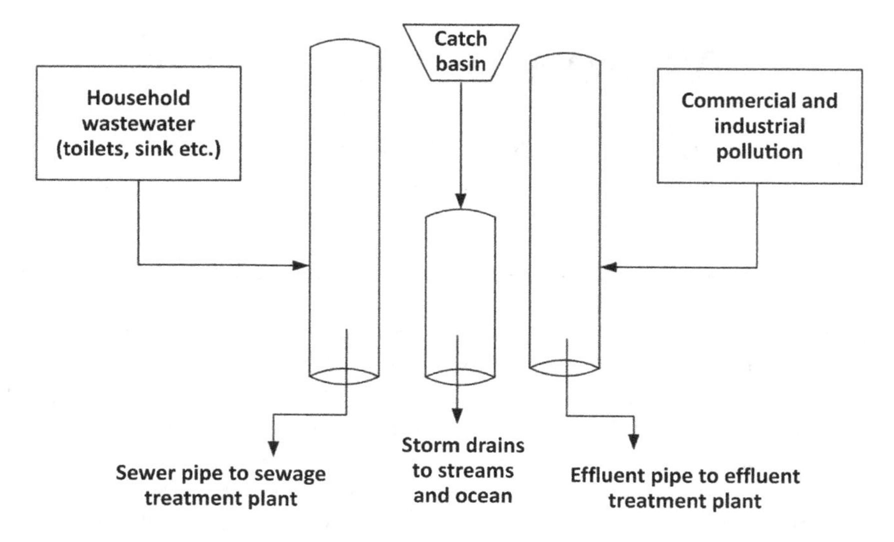

FIGURE 2.1 Overview of sewerage network in India

1. Greywater and/or septic tank effluent generated from various household activities.
2. Stormwater; that is, water which runs off the land and houses as a result of rainfall.

In rural areas, open channels are present which can be easily converted or improved to covered surface drains. The primary purpose of covered surface/stormwater drains is to collect and channelize the wastewater/rainwater from the individual houses/residential areas in a clean and well-organized manner. It is a necessary practice to collect, transport and treat in a manner such that there are no negative effects on public health and the environment, and also that do not cause further problems to residents and damage to other elements of infrastructure.

Two types of channels – (a) half-rounded central channel; and (b) outer channel – are shown in typical sections of covered surface drain (Figure 2.2a and Figure 2.2b). The half-rounded central channel facilitates the peak flow of wastewater in dry season, while the outer channel helps in the smooth discharge of stormwater in rainy season. It is also recommended that the floor of outer channel rather be inclined downward to the central channel. Meanwhile, it is observed that open drain/channel faces a greater friction than a pipe; laying down the pipe into the open channel and covering it is an alternative approach to achieve better pipe flow in relatively flat areas.

Small-bore sewers

Covered surface drains are low-cost and best suited for the collection of greywater/stormwater, but blackwater or mixed with greywater requires a more appropriate

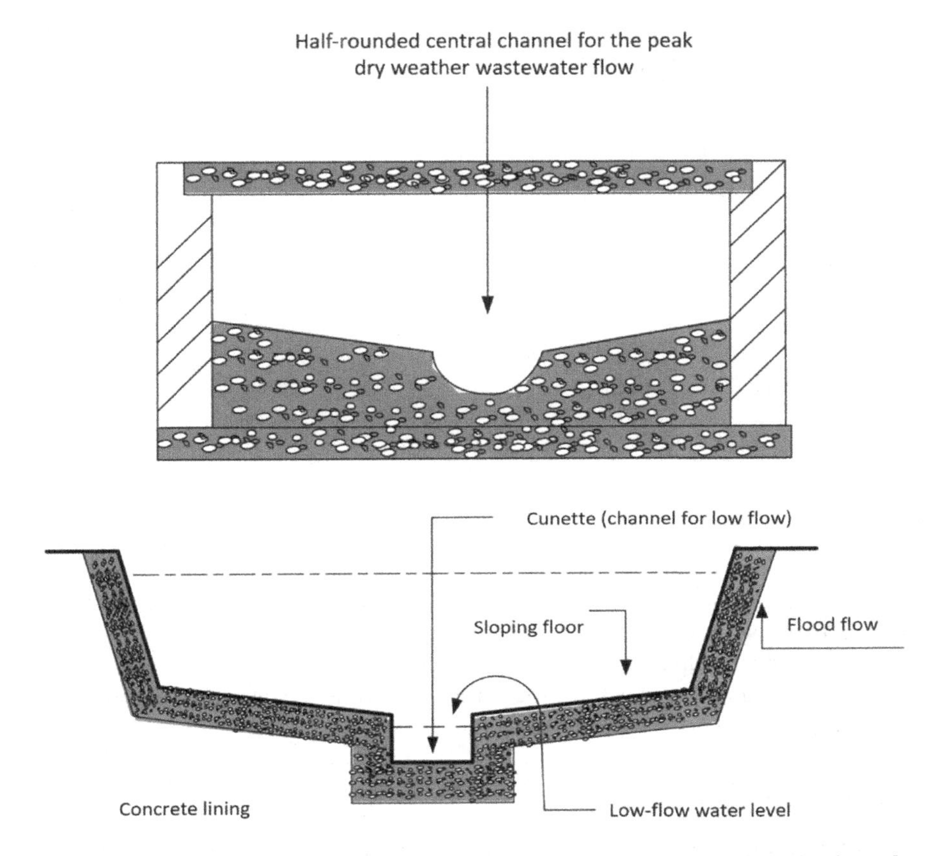

FIGURE 2.2 Covered surface drains: (a) half-rounded central channel; (b) outer channel

and sustainable approach such as small-bore/shallow sewer for the collection of wastewater. Small-bore sewer systems are dedicated to collect only the liquid portion of domestic wastewater and transport it to off-site systems for proper treatment and disposal. To screen out the grit, grease and floating materials from the wastewater flow, interceptor tanks (like septic tanks) are installed after individual houses or a particular residential area. The size of interceptor tanks depends upon the volume and inflow of wastewater; they should be regularly cleaned to remove the settled solids. A typical small-bore sewer system includes the following features (Figure 2.3).

1 *House connections*, which are built at the inlet to the interceptor tank.
2 *Interceptor tanks*, which are designed to keep the liquid flow up to 12–24 hours and to remove the solids from the wastewater flow.
3 *Sewers*, which includes small-bore pipes buried into the ground at a definite depth sufficient to collect the settled wastewater from most connections by gravity.

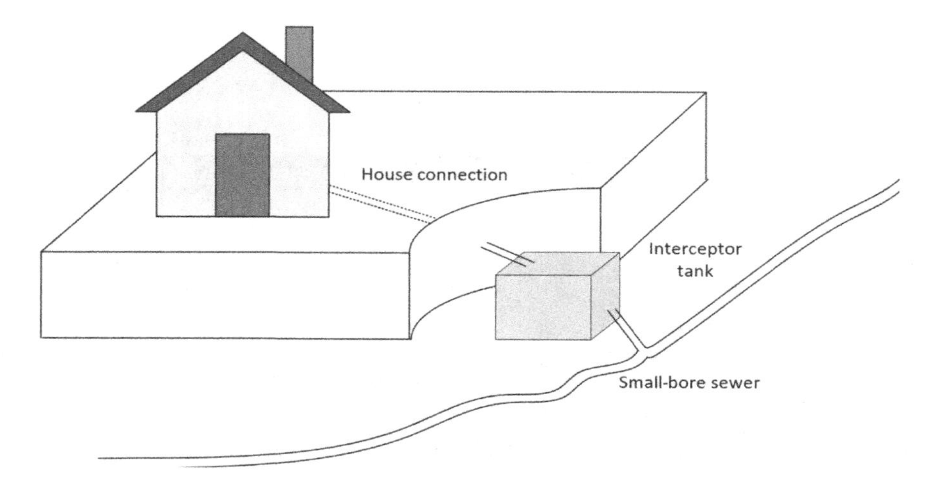

FIGURE 2.3 Small-bore sewers

4 *Cleanouts and manholes*, which are designed to allow access to sewers for regular inspection and maintenance.
5 *Vents*, which help to maintain free-flowing conditions for the wastewater.
6 *Pumping stations*, which lift the wastewater from one elevation to another station.

Conventional/simplified sewers

Conventional/simplified sewers are the best and most comprehensive solution for collection of wastewater in the areas where large volumes of wastewater are generated. These sewerage systems involve the collection and transportation of greywater and blackwater from the houses to the treatment plant for their treatment and disposal. High cost is involved in construction of conventional/simplified sewers; further, these sewer systems are primarily built deep inside the soil according the topographical profile of a particular area.

Simplified sewer systems can be implemented instead of conventional sewer systems. The simplified sewer seems to be the same as the conventional sewer system but needs reduced design criteria such as smaller diameters and slopes as compared to conventional sewer systems. Due to this, these systems are low-cost alternative to the conventional sewer system. These sewer systems also built shallow and constructed at locations that are easily accessible but not under heavy traffic (cars, etc.).

ON-SITE WASTEWATER TREATMENT SYSTEMS

Grease traps

Grease traps are the chambers which are used to trap oil and grease from wastewater, especially from kitchens. They are made out of plastic, metal, concrete or

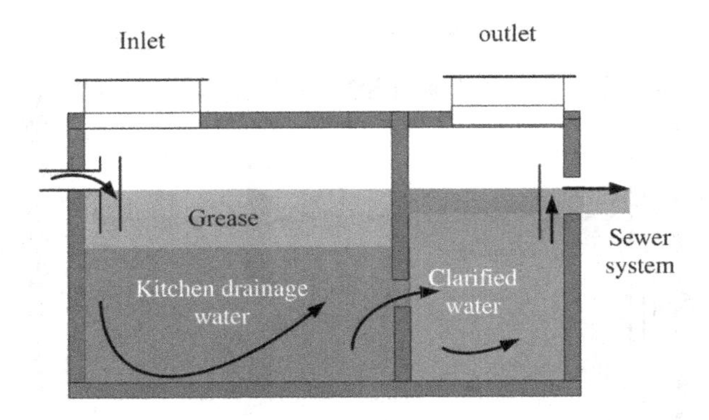

FIGURE 2.4 Grease trap

brickwork with an odor-tight lid. To prevent turbulence at the water surface and separate floating components from the effluent, baffles or tees are used at the inlet and outlet of the chamber (Figure 2.4). Grease traps are either fitted directly under the sink or a larger grease interceptor is installed outdoors for the collection of larger amounts of oil and grease. The under-the-sink grease trap is relatively cheap, but needs frequent cleaning (once a week or month), whereas larger grease interceptors are constructed to contain much more oil and grease for a longer time (6–12 months), but require a higher capital cost. Further, grease traps can be designed along the concept of septic tanks to contain grit and other settleable solids through sedimentation.

Septic tanks

Septic tanks are underground water-sealed tanks either constructed or fitted with pre-casted materials which receive and partially treat raw domestic blackwater. The heavy solids sink to the bottom of the tank while greases and lighter solids tend to float at the top; wastewater is either conveyed to the next tank or pit (in case of two-pit systems) or sent to the sewer network for further treatment and discharged to soak pits or leach pits (Figure 2.5).

These systems are one form of improved sanitation method, implemented to collect, store and treat the black wastewater; plays an important role in fecal sludge management (Strande et al., 2014; Shaw and Dorea, 2021). On the principles of ISO (2018), septic tanks minimize resource consumption (*e.g.* water, energy) and convert human waste to a safe output. They could be appropriate standalone sanitation systems in the area, which lacks conventional sewer system or it is unfeasible to build these systems. Further they are the most adaptable systems worldwide for the household sewage management and proved to capable to fulfill basic sanitation needs and provide social and environmental sustainability (Withers et al., 2014). It is reported that 3.1 billion peoples are using improved on-site sanitation systems in which 1.5 billion peoples opted for septic tank systems (UNICEF, 2019).

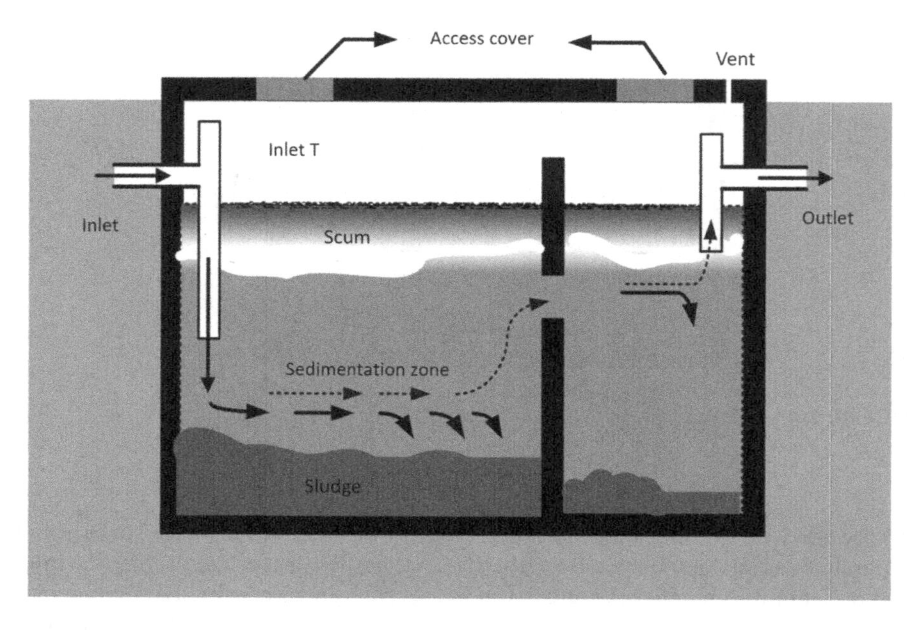

FIGURE 2.5 Septic tank

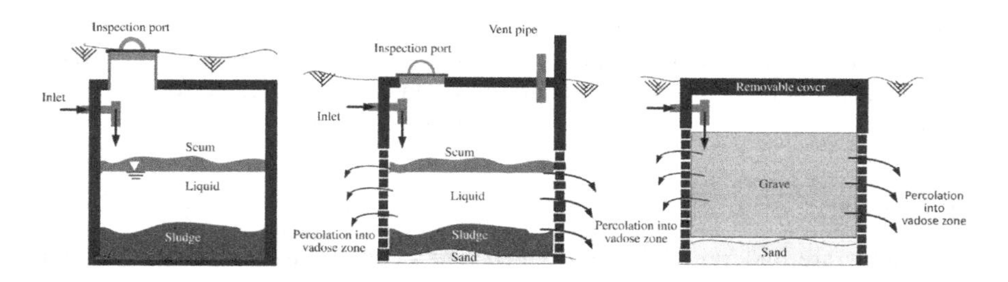

FIGURE 2.6 Soak pit or leach pit

Soak pits or leach pits

Soak pits or leach pits are the simplest method for the on-site partial treatment of the pre-settled blackwater effluent from a septic tanks or centralized treatment facilities, which is discharged to an underground perforated pit from which it can infiltrate to the surrounding soil (Adegoke and Stenstrom, 2019). As the blackwater percolates from the pit to the soil, small particles are filtered out through the soil matrix, while microbial activity degrades the organic matter in the blackwater (Figure 2.6). These soak pits have proven to be ultimate solution for isolated areas, and best suited for soil with good absorptive properties but not appropriate for clay, hard-packed or rocky soil. Soak pits are cost-effective and easy to build. They are underground structures and are designed for absorption of all wastewater produced by a household, leading to the groundwater recharging. As wastewater percolates through the layers of the soil, there is always the risk of groundwater contamination, especially in the locations with high groundwater table (Mondal, 2014; Adegoke and Stenstrom, 2019). Further, the working life of soak pits or leach pits substantially decreases without provision of a Nahani device or a screening medium to split out heavy solids.

REFERENCES

Adegoke, A. A., and T. Stenstrom. 2019. "Cesspits and Soakpits." In *Global Water Pathogen Project*, edited by J. B. Rose and B. Jiménez Cisneros. Michigan University, E. Lansing, MI: UNESCO. https://doi.org/10.14321/waterpathogens.58.

Carr, R., and M. Strauss. 2001. "Excreta-Related Infections and the Role of Sanitation in the Control of Transmission." *Water Qual Guidelines Standards and Health*: 89–113.

ISO. 2018. ISO 30500:2018-Non-sewered Sanitation Systems – Prefabricated Integrated Treatment Units – General Safety and Performance Requirements for Design and Testing. Reference number: ISO 30500:2018(E).

Mehta, V. K., R. Goswami, E. Kemp-Benedict, S. Muddu, and D. Malghan. 2013. "Social Ecology of Domestic Water Use in Bangalore." *Economic and Political Weekly*, 15: 40–50.

Ministry of Jal Shakti, Government of India. 2020. *National Annual Rural Sanitation Survey (NARSS) Round-3 (2019–20): National Report*. New Delhi: Government of India.

Mondal, P., A. Nandan, N. A. Siddiqui, and B. P. Yadav. 2014. "Impact of Soak Pit on Groundwater Table." *Environmental Pollution & Control Journal* 18: 12–7.

Shaw, K., and C. C. Dorea. 2021. "Biodegradation Mechanisms and Functional Microbiology in Conventional Septic Tanks: A Systematic Review and Meta-Analysis." *Environmental Science: Water Research & Technology* 7, no. 1: 144–55. https://doi.org/10.1039/D0EW00795A.

Strande, L., and D. Brdjanovic. 2014. *Faecal Sludge Management: Systems Approach for Implementation and Operation*. London, UK: IWA Publishing.

UNICEF. 2019. *The State of the World's Children 2019. Children, Food and Nutrition: Growing Well in a Changing World*. New York: UNICEF.

Withers, P. J., P. Jordan, L. May, H. P. Jarvie, and N. E. Deal. 2014. "Do Septic Tank Systems Pose a Hidden Threat to Water Quality?" *Frontiers in Ecology & the Environment* 12, no. 2: 123–30. https://doi.org/10.1890/130131.

3 Wastewater treatment
Decentralized systems

Wastewater treatment usually needs different stages that can vary depending on the treatment required for a particular type of wastewater. These stages are designated as preliminary, primary, secondary and tertiary (mainly disinfection or advanced treatment). The different processes or methods that are used in wastewater treatment classified into three main categories: physical, chemical and biological processes. Some of these processes are executed together in a single treatment plant according the required treatment of wastewater.

PHYSICAL PROCESSES

Physical processes include screening, flow-equalization, sedimentation, aeration and clarification. Screening involves a bar screen with uniformly sized openings placed in the flow of wastewater, leading to the removal of large solids such as rags, plastic and foam particles that can clog or damage the pumps and machines used in later processes. Most of the treatment plants are planned for a nearly constant flow and amount of wastewater with a steep range of fluctuation. To ensure that there is a process of flow equalization, whereby wastewater is collected and stored in equalization tanks, allow it to mix to get a homogenous quality; wastewater is then pumped to the next treatment steps at a constant rate. Further, there is need to calculate the hourly variation of the flow of wastewater to calculate the required volume of an equalization tank.

In the sedimentation tank or settling pond or primary clarifier, flocculation (process of colliding of the particles to form flocs) occurs in mixed uniform wastewater, leading to settling of flocs by gravity settling. Through this, sand and suspended solids are settled to generate clarified wastewater and sludge from the physical treatment. To facilitate the growth of micro-organisms, aeration (by oxygen supply for respiration) provided to clarified wastewater leads to microbial degradation of organic matter (this is described in more detail in the upcoming section on biological unit processes). Aeration is facilitated either mechanically or by allowing exposure to atmospheric air or providing air through diffusers.

CHEMICAL UNIT PROCESSES

Chemicals are usually added to the clarified wastewater to change qualities like pH control, chemical precipitation, coagulation and oxidation. Industrial wastewater rarely has an acidic or alkaline pH that needs to be adjusted to make the wastewater neutral in the range of 6–8 pH, favoring suitable microbial growth in biological treatment. For pH neutralization, two types of chemicals are

DOI: 10.1201/9781003407690-3

used: (a) sodium hydroxide, sodium carbonate, calcium carbonate or calcium hydroxide for acidic waste (low pH); and (b) sulfuric acid or hydrochloric acid for alkaline waste (high pH). Coagulation is the aggregation of minute solid particles dispersed in wastewater into a larger mass. For this, coagulants such as aluminum sulfate (alum) or ferrous sulfate can be used to aggregate minute particles in to a larger masses called flocs. Chemical coagulation–assisted flocculation can be accelerated by the use and mixing of polyelectrolyte, a flocculent, in which particles aggregate together to form flocs that led to quick settling.

BIOLOGICAL UNIT PROCESSES

The wastewater contains varied amount of organic and inorganic compounds which need to be removed or reduced through the use of micro-organisms, mainly bacteria. These micro-organisms tend to break down complex compounds into simpler and less toxic compounds and use these compounds for their growth and proliferation. These micro-organisms can be utilized through two ways: (a) suspended microbial growth (e.g., activated sludge); or (b) attached microbial growth (e.g., solid film). In both ways, micro-organisms (in large numbers of colonies) are mixed and retained into tank or reactor (secondary clarifier) containing clarified wastewater with a supply of excess oxygen.

In suspended growth systems, micro-organisms proliferate in small aggregates or "flocs" termed as activated sludge. This leads to more treated wastewater and activated sludge settling in the tank or reactor; from there, a part of it is conveyed to the aeration process to recover bacterial population for continuous biological treatment and the remainder is sent to a sludge handling unit. In attached growth systems, an inert support medium provided to the micro-organisms to grow as thin layer or "bio-film") on the surface of that medium acts as a percolating or biological filter. The support medium could be small pebbles or a plastic support which allows formation of the bio-film. The wastewater is traditionally either sprayed onto the support medium to percolate through a packed bed with oxygen inflowing from the air; or, in recent designs, the support medium (usually plastic) is immersed in the wastewater and air is supplied from the bottom of the tank or reactor. In comparison to traditional biological filters, advanced submerged fixed-film (plastic) filters need less land but require a final clarifying agent to remove bio-film particles that detach from the medium which not need to be recycled for the reactor.

The treated wastewater undergoes further treatment like disinfection or advanced processes, or it is discharged into the environment; which one mainly depends on the required quality or use of wastewater for various applications. The treated wastewater still has the problem of color, dissolved solids and toxic compounds that are removed by the advanced treatment. A number of methods, mainly filtration, are utilized in advanced treatment. Finally, treated wastewater involves disinfection such as chlorination and ozonation that kills the microbes, especially pathogenic microbes presented in the wastewater.

SLUDGE MANAGEMENT

The excess activated sludge after recycling is conveyed to a sludge handling or processing unit for proper disposal so that the pollutants in this sludge do not contaminate or pollute the environment. Sludge handling processing involves the conversion of compost, ashes or other end products. The characteristics of the end products primarily depend up on sludge type (primary, secondary or mixed), their treatment and the processes used for the treatment.

WASTEWATER TREATMENT TECHNOLOGIES

The technologies that are employed for the wastewater treatment in India, can be classified in to the following two broad categories.

1. Conventional technologies
2. Advanced technologies

CONVENTIONAL TECHNOLOGIES

Conventional technologies are easy and simple in their installation, and their operation does not need advanced machinery and equipment. The following technologies are utilized in wastewater treatment plants.

- *Activated sludge process (ASP):* This is an aerobic method in which oxygen is supplied to the wastewater to produce a biological flake. These flakes are removed as sludge, resulted in the reduction of organic content in the wastewater (Haydar et al., 2007). BOD and total suspended solids are significantly removed through this ASP method, which also facilitates biological nitrification and denitrification in the wastewater.
- *Extended aeration:* This is a modified form of ASP, whereby oxygen is supplied for longer periods to improve aerobic digestion of the organic content through endogenous respiration. It promotes the faster stabilization of the organic matter under aerobic conditions and also the elimination of the gaseous end products into the air (Sotirakou et al., 1999).
- *Trickling filters:* These are comprised of a basin or tower filled with support media such as stones or plastic or wooden slats. These support media provide a surface for the micro-organisms to attach to and develop a biological layer or fixed film. This layer or film acts as a filter to retain and stabilize the organic matter in the wastewater.
- *Karnal technology:* This is an emerging technology in which trees are planted on ridges 50 cm high and placed 1 meter from each other. The untreated wastewater discharges in the furrows between the planted trees and percolates in the soil; from there, nutrients are absorbed by the soil and plants. It is necessary to maintain the time interval of supplying wastewater,

which could be 12–18 hours, so that there is no standing water left in the trenches.

- *Upflow anaerobic sludge blankets:* UASBs are an anaerobic method, particularly appropriate for wastewater that contains soluble pollutants and particulate matter. As the UASB process does not require mechanical components or external energy, there is no need for electric power to run it. During the process, sludge and biogas of high calorific value is produced as by-products (Bal and Dhagat, 2001).
- *Oxidation ponds:* These are designed to provide secondary treatment to the wastewater. In this method, heterotrophic bacteria play an important role in the degradation of organic matter present in the wastewater. Oxidation ponds further facilitate the growth of algae in the wastewater, due to the abundance of cellular material and minerals (Ho and Goethals, 2020). This results in the faster decomposition of organic matter by heterotrophic bacteria because the presence of more oxygen in the wastewater produced by the algae.
- *Waste stabilization ponds:* WSPs are shallow artificial ponds consisting of one or more in a series of anaerobic, facultative and maturation ponds. The anaerobic ponds act as primary treatment, where suspended solids and soluble content of the organic matter are removed (Coggins et al., 2019). Wastewater is then conveyed to facultative ponds as during the secondary treatment to remove most of the remaining BOD through algae and heterotrophic bacteria. For tertiary treatment, maturation ponds are utilized to remove the pathogens and inorganic nutrients.
- *Aerated lagoons:* These are simply earthen basins having an inlet at one end and an outlet at the other end to allow wastewater to drain way from these structures. In this method, aeration usually takes place in a mechanical manner, which promotes the growth of micro-organisms to degrade the organic matter in the basin (Nunes et al., 2021).

Currently, ASP is the most widely used wastewater technology in India. Other technologies used prominently methods are the UASB technique and oxidation ponds. The least used are the extended aeration, trickling filter and Karnal technology systems.

Advanced technologies

In an effort to improve the quality of treated wastewater and improve the efficiency of operations and maintenance, several municipalities have begun to adopt advanced wastewater treatment technologies. The following are the most commonly used technologies for wastewater treatment.

- *Sequencing batch reactor:* SBR is a modified version of the conventional ASP method Fernandes et al., 2013). It involves a batch process that includes a sequence of primary settling, aeration, secondary settling and decanting of the treated wastewater in the same tank in a time-bound manner. This method is capable of not only removing most of the BOD, but also removing nitrogen and phosphorous simultaneously. It does not need secondary clarifiers, sludge pumping stations, etc.

- *Moving bed bioreactor/fluidized aerobic bioreactor:* The MBBR method is quite like the ASP process, but it permits suspension of media in the reactor to favor the growth of microbes (Bassin and Dezotti, 2018). Therefore, microbial growth is maximized in the aeration tank compared to the conventional aeration process. The FAB is quite like the MBBR process, but it uses stationary media in the reactor to favor the growth of microbes and stationary media to be fluidized in the aeration tank (Singare, 2019).
- *Membrane bioreactors:* MBRs include the aeration process and secondary clarification process in the same tank by allowing the aerated mixed wastewater to flow through membranes. In this method, contaminants are separated in downstream tank instead of settling it as like in most of treatment methods (Obotey Ezugbe and Rathilal, 2020). MBR technology promises the best quality treated wastewater through removing all BOD content and suspended solids. Further, it requires less power as it operates under low suction. MBR is a less frequently used treatment technology in the country, as compared to SBR and MBBR/FAB technologies.

Emerging technologies

In addition to conventional and advanced treatment technologies, more emerging wastewater treatment techniques are being implemented in the country. These include biological filtration and oxygenated reactor (BIOFOR), high rate activated sludge BIOFOR-F technology, submerged aeration fixed-film (SAFF) technology, fixed bed bio-film ASP (FBAS), rim flow sludge suction clarifiers/bio-towers and eco-bio blocks. As compared to conventional and advanced technologies, these technologies offer a series of important advantages – less space requirement, ease of operation and maintenance, absence of odor and aerosol during the treatment process, etc.

- *BIOFOR:* BIOFOR includes an improved primary treatment by using the coagulants and flocculants, and a two-stage filtration process through a biologically active media with improved external aeration (Sharma and Singh, 2013).
- *High rate activated sludge BIOFOR-F:* This is an enhanced mechanized technology that excludes a primary sedimentation process in the flow scheme. It includes aeration and rapid sand filtration heating up the digester, and follows the temperature-controlled anaerobic sludge digestion.
- *SAFF:* This method includes two-stage biological oxidation, in which fixed-film media – unconventionally plastic media – is used with improved oxygen supply through submerged aeration (Gurjar et al., 2019). This process leads to higher organic matter removal through a large biomass and solid retention time.

- *FBAS:* This method basically follows the ASP process, but it involves plant roots as a media to grow and develop the fixed bio-film. FBAS comprises series of biological reactors in which fixed bio-film is sustained with a separate aeration system at each step of the process. The plant roots and additional textile media act as bio-film carriers and help with the biodegradation of contaminants in the wastewater.
- *Bio-towers:* Bio-towers are basically clarifiers which have inlets along the edges. They have suction boxed arms instead of common scrapers to suction out the produced sludge from the floor (Doelle et al., 2020).
- *Eco-bio blocks:* Eco-bio blocks are exfoliated bricks, made up of volcanic ash with the property of resistance to decomposition. These bricks facilitate the microbes to grow well into their crevices. The microbes in the bricks perform aerobic, anaerobic or facultative activity, depending on the prevailing oxygen or septic conditions Luong et al., 2017). This method is typically utilized in "polishing" of the treated wastewater before it is discharge into water bodies.

TABLE 3.1

Cost comparison of various technologies used in sewage treatment plants

Technology	Average capital cost (secondary treatment), Million Rs/MLD	Average capital cost (tertiary treatment), Lac Rs/MLD	Total capital cost (secondary + tertiary), million Rs/MLD	Total area(m^2) per MLD secondary + tertiary treatment	Average total daily power requirement kWh/day/ MLD	Total annual O&M costs, Lac Rs/MLD. Up to secondary treatment
Activated Sludge process	6.80	40	10.8	1000	185.7	353.02
Moving bed biological reactor	6.80	40	10.8	550	223.7	372.11
Sequential batch reactor	7.5	40	11.5	550	153.7	288.15
Upflow anaerobic sludge blanket + extended aeration	6.8	40	10.8	1100	125.7	290.72
Membrane Bioreactor	30	-	30	450	302.5	-
Waste Stabilization ponds	2.3	40	6.3	6100	5.7	116.09

Source CPCB (2013); Kumar and Tortajada (2020)

STATUS AND COST OF TREATMENT TECHNOLOGIES

In the Class-I cities, the most implemented treatment technologies are ASP with 59.5% of total installed capacity, and UASB with 26%. Along with these technologies, WSPs are also implemented in 28% of the treatment plants, which accounts for 5.6% of their combined capacity (CPCB, 2013; Kumar and Tortajada, 2020). In Class-II towns, the implemented treatment technologies are WSPs with 71.9% (72.4% of STPs) and UASB, with 10.6% (10.3% of STPs) of total installed capacity (CPCB, 2013; Kumar and Tortajada, 2020). ASP technologies are most suited and appropriate for large cities as they need less space than UASB and WSP technologies (Table 3.1).

REFERENCES

Bal, A. S., and N. N. Dhagat. 2001. "Upflow Anaerobic Sludge Blanket Reactor a Review." *Indian Journal of Environmental Health* 43, no. 2: 1–82.

Bassin, J. P., and M. Dezotti. 2018. "Moving Bed Biofilm Reactor (MBBR)." In *Advanced Biological Processes for Wastewater Treatment*: 37–74. Cham: Springer.

CPCB. 2013. *Performance Evaluation of Sewage Treatment Plants under NRCD*. New Delhi: Government of India.

Coggins, L. X., N. D. Crosbie, and A. Ghadouani. 2019. "The Small, the Big, and the Beautiful: Emerging Challenges and Opportunities for Waste Stabilization Ponds in Australia." *WIREs Water* 6, no. 6: e1383. https://doi.org/10.1002/wat2.1383.

Doelle, K., Y. Qin, and Q. Wang. 2020. "Bio-tower Application for Wastewater Treatment." *Journal of Engineering Research & Reports* 11, no. 1: 1–7. https://doi.org/10.9734/jerr/2020/v11i117048.

Fernandes, Heloísa, Mariele K. Jungles, Heike Hoffmann, Regina V. Antonio, and Rejane H. R. Costa. 2013. "Full-Scale Sequencing Batch Reactor (SBR) for Domestic Wastewater: Performance and Diversity of Microbial Communities." *Bioresource Technology* 132: 262–8. https://doi.org/10.1016/j.biortech.2013.01.027.

Gurjar, Rishi, Akshay D. Shende, and Girish R. Pophali. 2019. "Treatment of Low Strength Wastewater Using Compact Submerged Aerobic Fixed Film (SAFF) Reactor Filled with High Specific Surface Area Synthetic Media." *Water Science & Technology* 80, no. 4: 737–46. https://doi.org/10.2166/wst.2019.316.

Haydar, S., J. A. Aziz, and M. S. Ahmad. 2007. "Biological Treatment of Tannery Wastewater Using Activated Sludge Process." *Pakistan Journal of Engineering and Applied Sciences* 1, no. 1: 1–6.

Ho, L., and P. L. M. Goethals. 2020. "Municipal Wastewater Treatment with Pond Technology: Historical Review and Future Outlook." *Ecological Engineering* 148: 105791. https://doi.org/10.1016/j.ecoleng.2020.105791.

Kumar, M. D., and C. Tortajada. 2020. "Wastewater Treatment Technologies and Costs." In *Assessing Wastewater Management in India*: 35–42. Singapore: Springer. https://doi.org/10.1007/978-981-15-2396-0_7.

Luong, H., T. Van Tuyen, T. T. Chinh, and D. T. Tu. 2017. "Application of the Modified EBB to Improve Lake Water Quality in Ha Noi." *Vietnam Journal of Science & Technology* 55, no. 4C: 186–91. https://doi.org/10.15625/2525-2518/55/4C/12150.

Nunes, J. V., M. W. B. da Silva, G. H. Couto, E. R. Bordin, W. A. Ramdsdorf, I. C. Flôr, V. A. Vicente, J. D. de Almeida, F. Celinski, and C. R. Xavier. 2021. "Microbiological Diversity in an Aerated Lagoon Treating Kraft Effluent." *BioResources* 16, no. 3: 5203–19. https://doi.org/10.15376/biores.16.3.5203-5219.

Obotey Ezugbe, Elorm, and Sudesh Rathilal. 2020. "Membrane Technologies in Wastewater Treatment: A Review." *Membranes* 10, no. 5: 89. https://doi.org/10.3390/membranes 10050089.

Sharma, C., and S. K. Singh. 2013. "Performance Evaluation of Sewage Treatment Plant Based on Advanced Aerobic Biological Filtration and Oxygenated Reactor (Biofor) Technology – a Case Study of Capital City-Delhi, India." *International Journal of Engineering Science Innovation and Technology* 2, no. 4: 435–42.

Singare, P. U. 2019. "Fluidized Aerobic Bio-reactor Technology in Treatment of Textile Effluent." *Journal of Environmental Chemical Engineering* 7, no. 1: 102899. https://doi.org/10.1016/j.jece.2019.102899.

Sotirakou, E., G. Kladitis, N. Diamantis, and H. Grigoropoulou. 1999. "Ammonia and Phosphorus Removal in Municipal Wastewater Treatment Plant with Extended Aeration. Global Nest." *International Journal* 1, no. 1: 47–53.

4 Energy problems in wastewater recycling

WATER–ENERGY NEXUS

Water supply needs energy starting from its extraction, transportation to areas where it is needed and re-supply and distribution, leading to the generation of huge amount of wastewater. Further, a significant amount of energy is utilized in the collection, treatment and utilization of water recycling. This interrelated and mutually constraining relationship between water and energy is known as water–energy nexus. To simplify this water–energy nexus, the Alliance to Save Energy coined the term Watergy, which provides us the opportunity to address energy and water issue at the same time, because conserving one leads to conservation of the other (Barry, 2007).

Through this Watergy approach, cities and town can save significant amounts of energy, water and financial costs by implementing some technical and administrative interventions in water supply and water recycling systems. Consistent supply of quality water can be ensured by using minimum quantities of water and energy. In coming years, this interdependency of water and energy could be more vulnerable and relevant, as water plays a crucial role in energy production and energy would be crucial to the development of the water sector. The Watergy concept relies on the simple basic fact that every liter of water that passes through a supply system consumes a significant amount of energy cost. Considering this reality, water supply and wastewater management are energy-intensive utilities, which can motivate the policy makers to make more energy-efficient and sustainable systems.

WATER–ENERGY NEXUS AND CLIMATE CHANGE

India is currently facing water shortages in most of its big cities, which could be further worsened by the devastating effects of climate change. Climate change may increase the frequency of floods, cyclones and droughts, and also be responsible for the mean sea level rise that could lead to the displacement of country's densely populated coastlines. Due to insufficient rain, energy demand will increase for the groundwater pumping and house/office cooling; this will lead to more carbon emissions, which would become contradictory to the commitment of keeping India's future per capita emissions below the level of already developed nations. This is the pledge of the Indian government as part of its Intended Nationally Determined Contribution submitted to the UN Framework Convention on Climate Change in October 2015 (UNFCCC, 2015). In this pledge, it is also promised that by the year 2030, emissions intensity of the country will reduce by 33%–35% of the previous level measured in 2005.

DOI: 10.1201/9781003407690-4

ENERGY EXPENDITURE IN WATER AND WASTEWATER MANAGEMENT

Due to old and inefficient technology, there is loss of significant amount of energy in water supply and wastewater treatment facilities (Table 4.1), which leads to increase in energy-related expenditures. The cost of energy expenditure in water supply is usually more than the expenditure associated with wages, operation and maintenance, treatment and other costs in India (Figure 4.3). According to *A Report on Seventh Electric Power Survey*, there are more than 12,000 MUs of

TABLE 4.1
Water and wastewater utility systems that use energy

Stage	Operation	Energy using system
Extraction	Deep well extraction	Submersible or shaft turbine deep-well pumping systems
	Extraction from a surface source	Horizontal or vertical centrifugal pumping systems
Water treatment	Chemical (disinfection and clarification)	Piston-type dosing pumps
	Physical (e.g., filtration and sedimentation)	Pumping systems, fans, agitators and centrifugal blowers
Piping between source and distribution network	Sending the drinking water to the distribution grid	Submersible or shaft turbine deep-well pumping systems, and horizontal or vertical centrifugal pumping systems
	Booster pumping	Horizontal or vertical centrifugal pumping systems used to increase pressure of water going into the distribution system or to pump water to a higher elevation
Distribution	Distribution to the end users	Horizontal or vertical
Storm and sanitary sewer system	Sewerage and drainage	centrifugal pumping systems
Wastewater treatment	Physical (e.g., screening and sedimentation)	Pumping systems, fans, centrifugal blowers
	Chemical (e.g. clarification, disinfection)	Piston-type dosing pumps
	Biological	Pumping systems, agitators, aerators, centrifugal blowers
Support systems	Support functions associated with the utility building(s)	Lighting systems and heating, ventilation and air conditioning (HVAC)

Source: Barry (2007)

electricity consumed by the public water works in India, accounting for 40%–60% of the total operating cost of the water supply (CEA, 2007). It is also suggested that each urban local body (ULB) can save a minimum of 25%–40% on energy expenditure through energy efficiency. According to the CSE's survey of 71 cities, the cost of electricity found to the highest contributor in the money spending on water utilities; it may be because water is pumped from distant sources and then wastewater is pumped out to the treatment facilities usually outside the city (Rohilla et al., 2017).

Apart from inefficiency, most Indian cities are facing huge distribution loss of water; it is revealed that there is an average of about 35% distribution loss which is equal to the loss of 6,877 MLD of water supply (CPCB, 2009). This loss of water (calculated for the year 2005) could be sufficient to take care of additional requirement of water in 2011. These substantial leakages or distribution losses could be responsible for the increase in the cost of the water supply by up to 11% in metro cities, 5% for Class-I cities and 4% for Class-II and Class-III cities.

It is estimated that roughly 80% of the water supplied to households is discharged as wastewater (CPCB. 2009). In 2015, there were about 61,754 MLD of total sewage generated in India, but only 37% or 22,963 MLD of this wastewater treated in sewage treatment plants (STPs) (CPCB, 2009). The problem of more wastewater generation and inability of treatment of this wastewater could be much worse, if the private and unofficial groundwater usage or losses in the distribution system could be added to these figures (Figure 4.1). In order to get a sustainable solution to the water/energy crisis, there should not be only focus on the supply side to increase the per capita availability, but also a need to focus on the consequently greater wastewater generation and treatment.

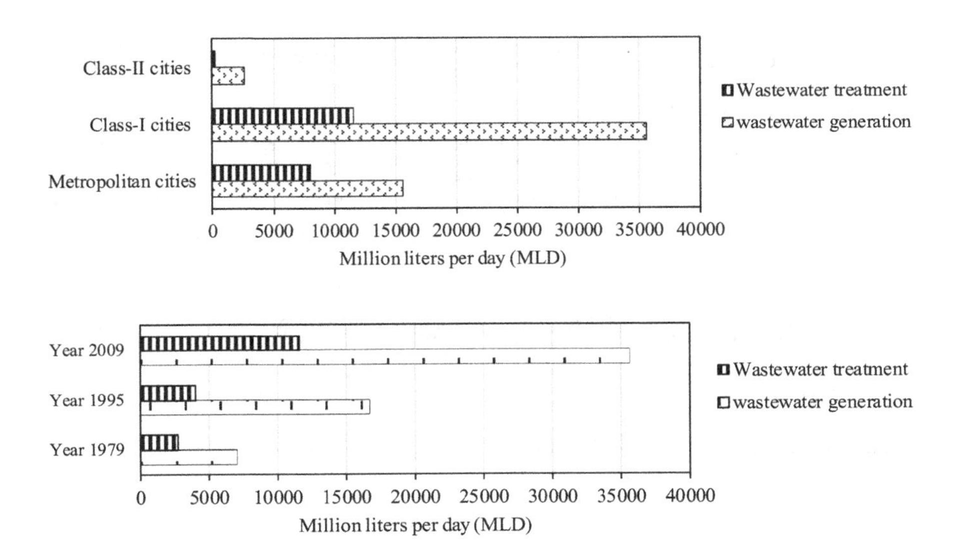

FIGURE 4.1 Water and wastewater generation in India

There is another area about non-revenue water (NRW) and associated water loss that generally is not included in the total water loss. It also includes the water loss from metering inaccuracies, unbilled consumption and unauthorized connections. Estimation of NRW is 16% in the United States, 15% in the United Kingdom, and 6% in the Netherlands, but it is much higher in India as Delhi – one of the better performing cities – has NRW of 59% (IIHS, 2014). Through implementing EE, Indian cities would be able to balance the water demand–supply gap (Figure 4.2).

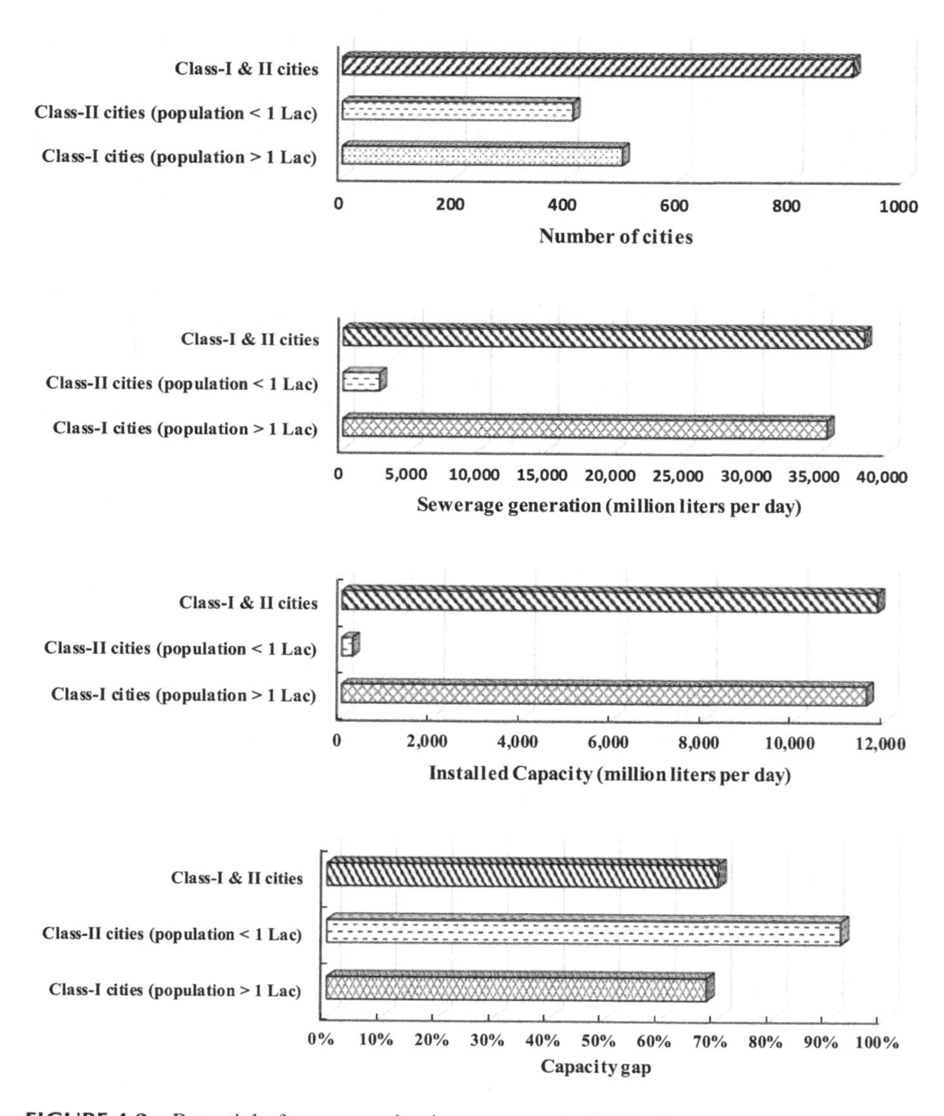

FIGURE 4.2 Potential of energy saving in water supply (IIHS, 2014)

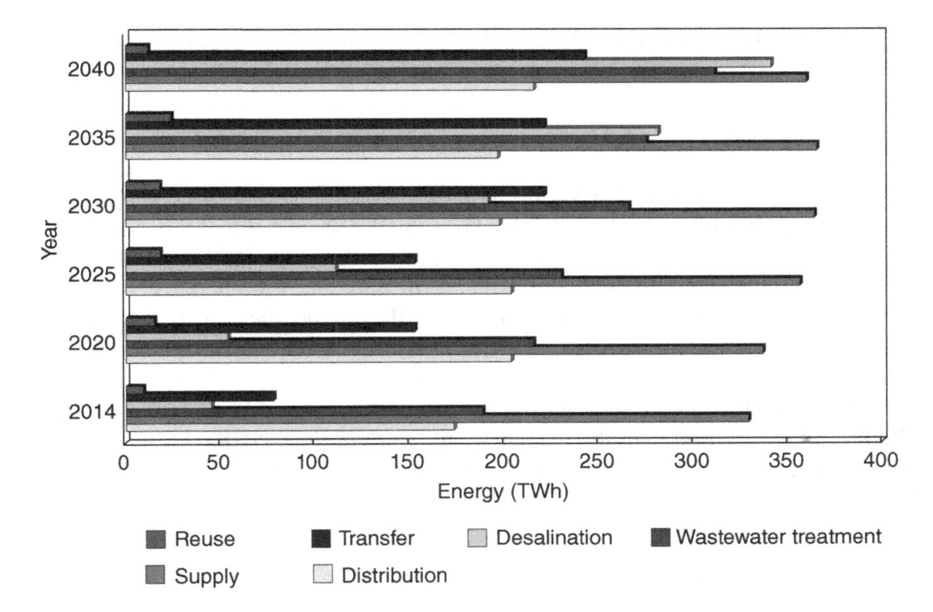

FIGURE 4.3 Electricity consumption in the water sector by process (IEA, 2016)

SOLUTIONS FOR ENERGY PROBLEMS

ENERGY EFFICIENCY (EE)

There is an estimation of nearly half of the world's population now residing in urban areas (municipalities), which accounts for the major portion of the water supply and wastewater management and associated energy consumption; therefore, the municipalities or ULBs should be in the focus to address the supply of water supply and associated energy consumption. Energy efficiency in water and wastewater systems could be achieved through the advancement of technology and science, and the use of modern electrical devices in these sectors including HVAC (heating, ventilation and air conditioning) and lighting systems. Beside these factors, pumping and aeration – which are responsible for most of the energy consumption in water and wastewater treatment plants – could be significant sectors in reducing energy consumption.

HEATING, VENTILATION AND AIR CONDITIONING

Older HVAC equipment in water and wastewater treatment plants can be upgraded with newer high-efficiency systems to help reduce energy consumption. According to the California Energy Commission (CEA, 2007), the air-source heat pumps, air conditioning systems and water pumps can reduce energy consumption by up to 30%–40%, as they are 10.5, 11.5 and 15.2 times more efficient than older systems. Further, new controllers and timers could be helpful in reducing the power used in heating or cooling – and even shutting down the system according to occupancy of the facility.

Improved lighting systems

Lighting systems generally account for 35%–45% of total energy consumption in an office building, and it could be reduced up to 30% by installing energy-efficient lighting systems (CEA, 2007). The upgradation of the light bulbs, lamps and sensor technologies could be better strategy in reducing the energy consumption related to improve lighting system. Smyth (2012) found that the replacement of T-12 lamps with T-8 lamps reduces energy consumption by 33%, equivalent to the annual saving of USD 12 per fixture. As compared to incandescent lamps, high-intensity discharge lamps are an energy-efficient option due to longer lifespan (16,000–24,000 hours) than the incandescent lamps (2,000–20,000 hours). Further, the remote-based switching (involving controllers and sensors) could save 25%–50% lighting usage compared to manual switching.

Pumping and aeration

Pumping and aeration equipment are the primary systems that are responsible for maximum energy consumption in the water supply and wastewater treatment. The pressure of increasing loads, these systems pending for proper optimization and their inability to reduce unnecessary aeration blower or pump capacity lead to the loss of energy and funds, and are also responsible for the higher carbon emissions over the life of the treatment facilities. Centre Public Health Environment and Engineering Organization (CPHEEO) is the agency that provides the guidelines in India for the designing of water supply and wastewater systems, which are most probably primarily established in relation to the analysis of demand forecasts. There are many instances when the pumping machinery facing the situation of mismatch in current head and flow requirements leads to the pump operating on reduced load conditions for many years. It is due to lesser water supply requirements during the initial years against the installed pump capacity which leads to the adoption of inefficient techniques such as throttling for the pump operations; this could be responsible for the water and energy loss, but it could be saved through equipment retrofitting and replacement of inadequate and old water and wastewater systems infrastructure.

In the case of water and wastewater treatment, aeration systems alone account for about 50%–60% of total energy consumption of the aeration commonly provided by the mechanical systems and subsurface systems (Mcgee, 1999). The mechanical systems, propellers and blades are utilized to suck the air from the atmosphere and later push it into the wastewater, while subsurface systems include the diffusers placed inside the wastewater to provide oxygen. In comparison of mechanical systems, subsurface systems are more energy efficient and are divided in to coarse- and fine-bubble mixing systems. Due to larger surface area through creating small oxygen bubbles and much slower upwards movement, fine-bubble aeration systems are more popular and are replacing the mechanical aeration or coarse-bubble systems (U.S. EPA, 1999).

Heat recovery

Heat recovery is the process of trapping the energy from liquids or gases and repurposing it for other functions instead of wasting it. Heat recovery requires a heat exchanger like air or water as the gaseous or liquid medium, which transfers the heat from one source to another. In the case of heat recovery from a gas, the heat from old heated stale air or the hot exhaust is utilized to warm the fresh air. In case of heat recovery from a liquid, the heat from discarded hot water is utilized to heat the new fresh water or fresh air. Newton (2011) estimated that 235 billion kWh energy could be recovered from hot water annually that is discarded down drains. With these methods, the heat energy trapped from the old air or exhausted air and discarded hot water can be transferred to warm the new fresh air, resulting in the conservation of the power used to heat the air. It is easier to recover heat in colder climates, as it requires the repeated ventilation of fresh air, especially in winter.

The municipal sector of India is the second largest municipal system in the world and is also responsible for the consumption of about 4% of the total electricity used in the country. The energy consumption by public water works in India was around 18,364 million units (MUs) in 2011–2012 and it would be projected at around 36,861 MUs for the year 2021–2022 (CEA, 2007). During the Eleventh plan, the Bureau of Energy Efficiency (BEE) focused on water, wastewater and street lighting as target sectors, and initiated a demand-side management initiative. The primary aim of this initiative is reduction in the electricity consumption, leading to increased cost savings and making the ULBs more energy efficient. To assess the potential of EE in ULBs of the country, situational surveys were conducted in ULBs; out of 175 ULBs, the detailed project reports of 134 were prepared based on a techno-commercial assessment. It is estimated that 120 MW of electricity can be saved through EE projects in the 134 ULBs. There is significant progress on street lighting sector through replacing sodium vapor lamps with LED bulbs, but still, several EE projects in the water and wastewater sector are not happening at a large scale.

Clean energy sources

On-site power generation is essential for energy reliability and sustainability. In comparison to conventional energy sources, renewable energy sources could be better alternatives to achieve energy security and reduced air emissions. There are many renewable energy technologies which are widely used, such as hydropower generators, cogeneration systems using biogas, wind turbines and solar panels. These renewable energy technologies need significant capital investment and skilled labor in installation – but once started, they produce more environmentally sustainable electricity and have significantly lower operational and maintenance costs than energy based on fossil fuels.

TABLE 4.2
Cost of one unit electricity of different resources (Kost et al., 2021)

Year 2021	PV rooftop (small)	PV rooftop (large)	PV ground (utility)	Wind onshore	Wind offshore	Biogas	Solid biomass	Lignite	Hard coal	CCGT	Gas turbine
€/MWh	84.1	72.1	44.1	61.15	96.8	128.55	112.75	128.6	155.35	104.25	202.1
₹/MWh	7.35	6.30	3.86	5.35	8.46	11.24	9.86	11.24	13.58	9.12	17.67

Solar energy is the most abundant and cleanest source of energy that can be utilized for various large-scale applications. It can be used in two ways: (a) solar heat, for direct heating and cooling the systems; and (b) solar electricity, for running the equipment like pumps and other applications. Thus, solar energy can play a significant role in the water and water recycling sector in order to provide clean and sustainable sources of energy, and also to help with reduction of carbon emissions (Table 4.2) (Argaw, 2003; Hongbin et al., 2002; Ren et al., 2012).

REFERENCES

Argaw, N. 2003. *Renewable Energy in Water and Wastewater Treatment Applications; Period of Performance: April 1, 2001 – September 1, 2001 (No. NREL/SR-500–30383)*. Golden, CO: National Renewable Energy Laboratory (National Renewable Energy Laboratory, Office of Energy Efficiency and Renewable Energy).

Barry, J. A. 2007. *WATERGY: Energy and Water Efficiency in Municipal Water Supply and Wastewater Treatment. Cost-Effective Savings of Water and Energy*. The Alliance to Save Energy. New York: U.S.AID.

CPCB. 2009. *Status of Water Supply, Wastewater Generation and Treatment in Class-I Cities and Class-II Towns of India*. Ministry of Environment and Forests, Government of India. New Delhi: CPCB.

CEA. 2007. *A Report on Seventh Electric Power Survey*. Ministry of Power Government of India. New Delhi: Central Electricity Authority.

Hongbin, L., L. Tao, and L. Yuncang. 2002. *Status of Solar Photocatalytic Wastewater Treatment*. China: Yunnan Normal University.

IEA. 2016. *World Energy Outlook 2016*. Paris: IEA. https://www.iea.org/reports/world-energy-outlook-2016.

IIHS. 2014. "Sustaining Policy Momentum Urban Water Supply & Sanitation in India IIHS RF." Bangalore: Paper on Water Supply and Sanitation.

Kost, C., S. Shammugam, V. Fluri, D. Peper, A. D. Memar, and T. Schlegl. 2021. *Levelized Cost of Electricity Renewable Energy Technologies*. Freiburg, Germany: Fraunhofer Institute for Solar Energy Systems ISE.

Newton, J. 2011. "Kent County's Heat-Recovery Energy System." Kent County. Web. 8 Oct. 2013.

Ren, Z., Z. Chen, L. A. Hou, W. Wang, K. Xiong, X. Xiao, and W. Zhang. 2012. "Design Investigation of a Solar Energy Heating System for Anaerobic Sewage Treatment." *Energy Procedia* 14: 255–9. https://doi.org/10.1016/j.egypro.2011.12.926.

Rohilla, Suresh Kumar, Mahreen Matto, Shivali Jainer, Mritunjay Kumar, and Chhavi Sharda. 2017. *Policy Paper on Water Efficiency and Conservation in Urban India*. New Delhi: Centre for Science and Environment.

Smyth, E. 2012. "T12 Phase-Out in July." Precision P2 Paragon. Echo, Web. 20 Nov. 2013.

UNFCCC. 2015. *India's Intended Nationally Determined Contribution*: 1–38. New Delhi: UNFCCC/INDC.

U.S. EPA. 1999. *The Benefits and Costs of the Clean Air Act 1990 to 2010*. Washington, DC: U.S. EPA.

5 Solar powered wastewater recycling (SPWR)

SOLAR ENERGY AND ITS UTILIZATION

The radiant light and heat of the sun is called solar energy. To harness this energy, various technologies have been developed such as photovoltaics, solar architecture, solar heating, solar thermal energy, molten salt power plants and artificial photosynthesis. There are two main categorizations of solar energy: solar thermal and solar electricity.

SOLAR THERMAL HEATING AND COOLING

SOLAR THERMAL HEATING

Solar thermal systems work by transferring the sunlight heat to fluids. This fluid can be utilized for preheating of the main heating and can also be used for water heating for domestic purposes. There are broadly two classifications of solar panels: solar tubes and flat plates. It is cheap to use the flat plates, but when it comes to domestic hot water, the solar tubes system is more efficient.

Flat plate collectors

These collectors use a dark flat plate to absorb the solar energy, with a transparent cover over it to reduce the heat loss while allowing the solar energy to pass through. Also, the heat from the absorber is removed by utilizing a heat-transport fluid (air, antifreeze or water), along with a heat-insulating backing.

Evacuated tube collectors or solar tubes

Instead of passing liquid directly through the tubes, heat pipes are utilized in the core of the evacuated tubes. These collectors comprise multiple evacuated glass tubes with an absorber plate merged to a heat pipe. The heated pipe conducts the heat from the hot end and transfers it to the fluid – typically propylene glycol – and spreads it to a domestic hot water tank heat exchanger and/or a hydronic space heating system. The vacuum surrounding the tube reduces the heat loss to the outside in the form of either conduction or convection heat loss; therefore, the resultant efficiency is more than the flat plate collectors, especially in colder conditions.

- *Low-temperature technologies:* solar space heating, solar water heating, solar pond and solar crop drying (working temperature < 70°C).

DOI: 10.1201/9781003407690-5

- *Medium-temperature technologies:* solar cooling, solar distillation and solar cooking (70°C < working temperature < 200°C).
- *High-temperature technologies:* solar thermal power generation technologies such as solar tower, parabolic troughs and parabolic dishes (working temperature > 200°C).

SOLAR THERMAL COOLING

Depending upon the type of energy utilized for the cooling, the solar cooling system is categorized in two main classifications: cooling systems based on solar electric and solar thermal. In solar thermal cooling systems, the solar collectors collect the solar energy which is further transformed to thermal energy to drive the solar thermal cooling system via absorption, adsorption and desiccant cycles.

Absorption systems

Absorption systems are the oldest refrigeration technologies. A thermal compressor is utilized in absorption, while in conventional vapor compression, mechanical compressor is used. Absorbers used in thermal cooling contain a rich solution (rich in coolant) and the generator contains a solution heat exchanger. The most frequently used refrigerant/absorbent pairs are water/lithium bromide (H_2O/LiBr) and ammonia/water (NH_3/H_2O). Water works as coolant (refrigerant) and LiBr works as absorbent; for the other pair, ammonia is refrigerant and water is absorbent.

Adsorption systems

Similarity exists in the absorption refrigeration cycle and adsorption refrigeration cycle. The difference lies in that in adsorption, the highly porous material absorbs the refrigerant on its internal surface, while in the absorption system, liquid solution absorbs the refrigerant. In the adsorption refrigeration cycle, the solid acts as the adsorbent while the absorption pair – like activated carbon-methanol, carbon-ammonia, silica gel water, activated carbon-ethanol and zeolite-water – works as refrigerant.

Desiccant systems

Desiccant materials are used in this system, as in these materials, air moisture content can be removed by sorption processes. All the desiccant materials are capable of attracting moisture with varying capacities. For the climates having high humidity, the desiccant cooling system is the most suitable option for thermal comfort. Moreover, the low temperature gain from solar energy can be used by renewable energy, waste heat, and cogeneration to drive the cooling cycle.

SOLAR ELECTRICITY/SOLAR PHOTOVOLTAICS OR SOLAR CELLS

Whenever the sunlight strikes the solar cell, the photons present in the sunlight get absorbed and raise the energy level of the electrons; thus, the electrons get free from their atomic cells. The electric field present in the solar cell due to the p–n junction (the boundary between two types of semiconductor material) forces the free electron

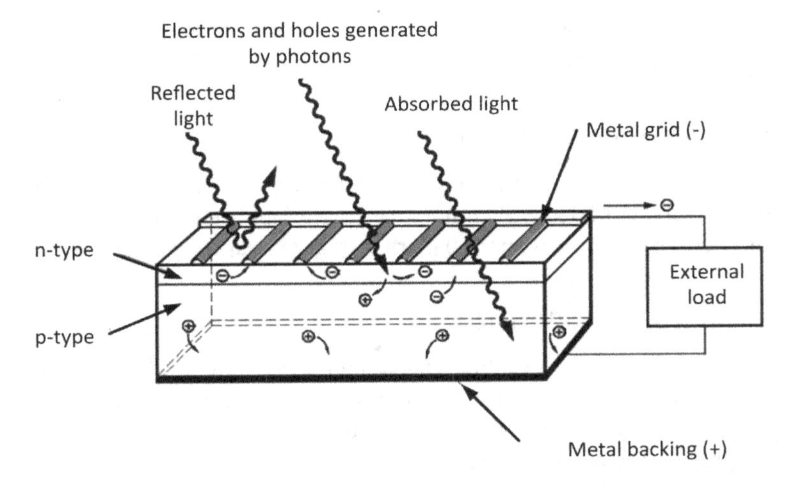

FIGURE 5.1 Working of a solar cell

into the n-region and the positive charges are collected to the p-region. The metal grid on the surface of the cell carries the negative charge while the positive charge is carried through the back plate (Figure 5.1) (Hammonds, 1998).

This phenomenon is called the photovoltaic effect, while the semiconductor device that transforms the solar energy to electricity is known as the photovoltaic cell. The number of solar cells can be assembled in series or parallel combinations to achieve the desired power level (Ayoub et al., 2009; Ibrahim et al., 2009). Power generation from the solar cell is uncertain and totally dependent upon the metrological conditions like the sun's geometric location, ambient temperature and irradiation levels (De Soto et al., 2006; Ikegami et al., 2001; Prajapati and Fernandez, 2020a).

The solar PV system includes a power generation unit that transforms the solar power into electricity by means of photovoltaics. Due to the modulus nature of the PV systems, the various ranges of capacities are available from a few to several tens of kilowatts (building-integrated systems or rooftop mounted), from residential to power stations of hundreds of megawatts. Nowadays, most PV systems are grid-connected, and few systems are of off-grid or standalone system configuration.

Rooftop solar PV systems

In these types of systems, the solar panel can be installed on the rooftop of a residential, commercial and industrial building (Prajapati and Fernandez, 2020b). The power thus generated is utilized to fulfill the load requirement.

In some cases, the power thus generated can be directly utilized for non-critical loads, without the need of battery backup, called *direct coupled solar PV systems*. The most common application of these systems is as water pumping systems. These systems pump water during daytime (sun is available). Thus, the total sunlight during day and the pump chosen determines the amount of water pumping. No storage is required as the system works when sunlight is available. As these systems involve the simplest configuration, the cost of these types of systems are comparatively low.

Battery-coupled solar PV systems

This system configuration involves the battery backup to maintain the continuity of power supply for the periods when no solar is available. The maintenance and the replacement cost get involve with the battery storage (Prajapati and Fernandez, 2021a).

Grid-connected solar PV systems

Instead of utilizing a battery backup, the system utilizes the grid power for periods when solar is unavailable. In most of the cases, the residential customer utilizes a rooftop solar PV system with a grid connection (Prajapati and Fernandez, 2019). Also, the grid power mostly generated with conventional fuels is nowadays typically installing renewable power for eco-friendly purposes (Prajapati and Fernandez, 2020c, 2021b). During the solar power production period, the power is consumed from the solar PV, and during night time, power can be purchase from the grid. In this way, the annual cost savings can be achieved by reducing grid power purchase (Prajapati and Fernandez, 2020d).

Hybrid solar PV systems

The solar PV system relies on an auxiliary source of power (fossil fuel generators or the grid). In this system, battery storage is utilized to avoid short-term fluctuations. This type of system is most commonly used for the critical applications or for places where large variations in sunlight occur throughout the year.

CONCENTRATING SOLAR THERMAL POWER (CSP)

CSP are the mirrors and tracking systems to capture and concentrate the spread sunlight into a small focused beam. The heat thus produced can be utilized as a heat source for various applications like for cooking, desalination of water and for conventional steam-based power plants. Among all the concentrating technologies, the most common and well-developed technologies are the solar power tower and the parabolic trough, while linear Fresnel reflectors and dish concentrators are the less well-developed technologies. To make the system more efficient, sun tracking and light focusing technologies are utilized. The thermal energy storage is most commonly used in the CSP to provide heat supply in the absence of sunlight. As the energy storage is quite efficient and cost-effective in CSP plants, almost all the CSP systems are in buildings with storage capacities of up to 15 hours.

BUILDING-INTEGRATED PHOTOVOLTAIC SYSTEMS

Building-integrated photovoltaic (BIPV) systems are cost-effective as the building materials hold the photovoltaic properties such as in roofing, glass and siding (Aaditya and Mani, 2018).

Instead of purchasing and installing the conventional material in construction, BIPV materials can be used for additional savings. Also, BIPV installations look architecturally more attractive when compared to roof-mounted PV structures. For small-scale electrical power production, BIPV systems can be used for the home, while for large-scale power generation, BIPV can be installed on large buildings. As the sun appears at lower angles in higher latitude locations, it is effective to utilize BIPV material for building.

SOLAR POWER GENERATION IN INDIA

India is blessed with rich solar energy potential of 5,000 trillion kWh per year; most parts of India receive radiations of about 35 kWh per square meter per day. According to the server of National Institute of Solar Energy, the overall energy potential of India is about 748 GW, while only 3% of the wasteland is covered by solar PV modules. This huge potential makes solar energy a main focus of India's National Action Plan on Climate Change with a National Solar Mission, which has launched on 11 January 2010. Until November 2020, the overall solar installed capacity of India was found to be 36.9 GW. Also, the government of India is committed to achieve a solar installation target of 100 GW by 2022; but due to Covid-19 pandemic and cost of solar panels, only 62 GW capacity developed so far. This target is aligned with Intended Nationally Determined Contributions (INDCs) of India, the project to ensure about 40% cumulative electric power installed capacity contributed from non-conventional energy resources and reducing the carbon emission of its GDP by 33%–35% of the 2005 level by 2030. Recently, India has surpassed Italy in solar power development and achieve the fifth position worldwide. A significant increment solar power capacity has been achieved which is more than 11 times in the last five-year period from March 2014 (2.6 GW) to July 2019 (30 GW). Thus, with the increased installation and technological development in solar power, solar tariff rates in India are quite competitive and many states have achieved grid parity.

Off-grid systems

Most of the solar power applications are off-grid applications, due to their remote location. These include terrestrial communication sites, remote homes, communication satellites, villages, and water pumps. Sometimes the hybrid system is also utilized in this system whereby an additional generator is required to charge the battery for the insufficient solar power periods. Although these types of systems becomes quite costly with the average cost of Rs 1 lakh with the additional battery installation that requires frequent maintenance. Under the decentralized solar PV mission, the objective of the government is to add on the solar power capacity of 118 MWp by 2020; which was successfully achieved with the various following application targets.

Street lighting systems: For the northeastern states and districts affected by left-wing extremism in remote having no facility of the street lighting, the government sets a target of installing 3,00,000 street lightening systems with grid connection.

Solar study lamps: Providing 25,00,000 solar study lamps to the northeastern states and districts affected by left-wing extremism in backward and remote areas to school students up to Secondary School Certificate level.

Off-grid solar power plants: These plants are installed for places where grid power is not available or is not reliable, installation has been done of individual size up to 25 kWp. Such plants mainly aim to provide electricity to hostels, schools, police stations, panchayats and other public service institutes.

Under the Pradhan Mantri Kisan Urja Suraksha evam Utthaan Mahabhiyaan (PK-KUSUM) scheme to provide energy to India's farmers, the government aims to add 25,750 MW solar capacity by 2022; but due to Covid-19 pandemic this scheme extended upto 2026 under three components: 10,000 MW ground-mounted decentralized renewable plants of individual size 2 MW, 17.5 lakh standalone solar powered agriculture pumps of individual pump capacity 7.5 HP and solarization of 10 lakh grid-connected agriculture pumps of individual pump capacity 7.5 HP.

Grid-connected systems

The direct current (DC) power from solar PV is converted to alternating current (AC) power through inverters and fed to the main distribution system for grid-connected applications. Grid-connected systems can provide the continuity of power supply during natural disasters by providing emergency power. As the renewable power is costlier as compared to conventional power, the use of grid-connected systems has increased. Costs for installation of grid-connected systems range between Rs 50,000 and Rs 75,000 per kWp depending upon the inverter and type of panel chosen (Table 5.1).

A plan of installing 100 GW of grid-connected solar power is targeted by 2022; but due to Covid-19 pandemic and cost of solar panels only 62 GW capacity developed so far. To achieve this target, the government of India has launched various schemes to promote solar power generation such as solar parks, a central public sector undertaking (CPSU) scheme, viability gap funding (VGF), a defense scheme,

TABLE 5.1
Description of costs

Cost of 1 kW rooftop solar system	1,00,000
Subsidy @ 30%	30,000
Cost after subsidy	70,000
Accelerated depreciation @ 80%	56,000
Tax rate @ 35% saved after Accelerated Depreciation (AD)	19,600
Net cost after subsidy and AD savings	50400

Source: Currency-Indian Rupees (INR)

canal top and canal bank schemes, a bundling scheme, and a grid-connected solar rooftop scheme.

India has achieved fifth rank (REN21, 2022) in the world in solar power; the capacity has increased 14 times in the last six years from 2.6 GW in March 2014 to 36.9 GW November 2020. India has achieved grid parity with technological improvement.

REFERENCES

Aaditya, G., and M. Mani. 2018. "BIPV: A Real-Time Building Performance Study for a Roof-Integrated Facility." *International Journal of Sustainable Energy* 37, no. 3: 249–67. https://doi.org/10.1080/14786451.2016.1261864.

Ayoub, N., R. Batres, and Y. Naka. 2009. "An Approach to Wicked Problems in Environmental Policy Making." *WSAES Transactions on Environment & Development* 5, no. 3: 229–39.

De Soto, W., S. A. Klein, and W. A. Beckman. 2006. "Improvement and Validation of a Model for Photovoltaic Array Performance." *Solar Energy* 80, no. 1: 78–88. https://doi.org/10.1016/j.solener.2005.06.010.

Hammonds, M. 1998. "Getting Power From the Sun." *Chemistry & Industry* 6: 219–22.

Ibrahim, M., K. Sopian, W. R. W. Daud, M. A. Alghoul, M. Yahya, M. Y. Sulaiman, and A. Zaharim. 2009. "Solar Chemical Heat Pump Drying System for Tropical Region." *WSEAS Transactions on Environment & Development* 5, no. 5: 404–13.

Ikegami, T., T. Maezono, F. Nakanishi, Y. Yamagata, and K. Ebihara. 2001. "Estimation of Equivalent Circuit Parameters of PV Module and Its Application to Optimal Operation of PV System." *Solar Energy Materials & Solar Cells* 67, no. 1–4: 389–95. https://doi.org/10.1016/S0927-0248(00)00307-X.

Prajapati, S., and E. Fernandez. 2019. "Rooftop Solar PV System for Commercial Office Buildings for EV Charging Load." In *IEEE International Conference on Smart Instrumentation, Measurement and Application (ICSIMA)*, Vol. 2019: 1–5. IEEE Publications. August. https://doi.org/10.1109/ICSIMA47653.2019.9057323.

Prajapati, S., and E. Fernandez. 2020a. "Performance Evaluation of Membership Function on Fuzzy Logic Model for Solar PV Array." In *IEEE International Conference on Computing, Power and Communication Technologies (GUCON)*, Vol. 2020: 609–13. IEEE Publications. October. https://doi.org/10.1109/GUCON48875.2020.9231202.

Prajapati, S., and E. Fernandez. 2020b. "Reducing Conventional Generation Dependency with Solar Pv Power: A Case of Delhi (India)." *Energy Sources Part A*: 1–12.

Prajapati, S., and E. Fernandez. 2020c. "Relative Comparison of Standalone Renewable Energy System Battery Storage Requirements for Residential, Industrial and Commercial Loads." *International Journal of Renewable Energy Technology* 11, no. 2: 111–25. https://doi.org/10.1504/IJRET.2020.108330.

Prajapati, S., and E. Fernandez. 2020d. "Solar PV Parking Lots to Maximize Charge Operator Profit for EV Charging with Minimum Grid Power Purchase." *Energy Sources, Part A*: 1–11. https://doi.org/10.1080/15567036.2020.1851325.

Prajapati, S., and E. Fernandez. 2021a. "Capacity Credit Estimation for Solar PV Installations in Conventional Generation: Impacts with and without Battery Storage." *Energy Sources, Part A* 43, no. 22: 2947–59. https://doi.org/10.1080/15567036.2019.1676326.

Prajapati, S., and E. Fernandez. 2021b. "Standalone Solar PV System for Grey Water Recycling Along with Electric Load for Domestic Application." *International Journal of Sustainable Engineering* 14, no. 5: 933–40. https://doi.org/10.1080/19397038.2020.1739168.

Renewables 2022 Global Status Report (Paris: REN21 Secretariat): *Global Status Report Supported by UNEP*. Paris, France.

6 SPWR for municipal wastewater

DOMESTIC OR MUNICIPAL WASTEWATER

Domestic or municipal wastewater comprises the total water generated from houses, apartments or societies and shops, restaurants or hotels; it contains mainly organic load with limited presence of heavy metals and chemicals. It is categorized in to two groups according to their origin: (a) greywater; or (b) blackwater.

Greywater includes the wastewater from showers, wash basins and clothes washing; sometimes it also includes the wastewater from kitchen sinks or dishwashers. As greywater does not include wastewater from lavatories, there is no presence of pathogens – but pathogens might be present from other sources like babies' nappies or diapers; still, pathogen concentrations are much lower than in wastewater from lavatories (Carr and Strauss, 2001; Adegoke and Stenstrom, 2019).

COMPOSITION AND CHARACTERISTICS OF GREYWATER

Greywater contains the residues of soap, shampoo, skin, hair, FOG and pathogens which are mixed from bathing, laundry and kitchen activities (which sometimes is not included in greywater). It always has a lower concentration of organic pollutants and nitrogen compounds than in the blackwater produced from the toilets. Further, there are many factors like quality of water supply, material of the water supply pipe and lifestyle of the households that are responsible for the physicochemical characteristics and composition of greywater.

It is reported that greywater from bathrooms alone make up about 50%–60% of total greywater; while all washing requirements account about 25%–30% including laundry sources that separately accounts for about 4% of total greywater (Loh and Coghlan, 2003; Poyyamoli et al., 2013). The greywater from kitchen sources accounts for around 10% of total greywater and includes contaminants such as food particles, oil and grease. Due to high amounts of organic contents, kitchen greywater is not considered as greywater by some researchers and it is advised to treat it separately using appropriate technology prior to reuse. It is also reported that the greywater from kitchens and dishwashers contributes nearly 50% of its COD requirement (Friedler, 2004).

GREYWATER RECYCLING

Collection of wastewater

Individual residential areas

In individual houses (including in industry complexes), raw greywater can be collected from baths, showers, laundries, hand basins and kitchens (if possible, collect

DOI: 10.1201/9781003407690-6

kitchen greywater separately or install a grease trap before storage) through drainage pipes to in-house storage tanks.

Shared residential areas

In apartment buildings and hostels, raw greywater can be collected from baths, laundries, hand basins and kitchens through shared drainage pipes. As greywater from kitchen contains significant amounts of oil and grease that can clog and decrease the efficiency of the filtration system, it can be possible to collect kitchen greywater though separate piped system; from these, oil and grease can be separated using grease traps before being discharged to the storage tank or conveyed to the greywater treatment facility.

Commercial areas

In commercial areas, greywater is generated from cleaning, hand basins and canteen areas – but as compared to residential areas, there is an absence of greywater from bath and laundry facilities and little or more presence of greywater from kitchens. They are subcategorized into: (a) commercial areas with little kitchen water like offices, institutions and shopping complexes etc.; and (b) commercial areas with more kitchen water like food shops, restaurants and hotels, etc.

Offices, institutions and shopping complexes

In these commercial establishments, greywater is mainly generated from hand basins and canteen areas. Greywater from hand basins is either provided to flush toilets or collected separately for the treatment. Canteens running in the daytime produce little greywater which contains oil and grease and detergents collected separately.

Food shops, restaurants and hotels

In these commercial establishments, greywater mainly generated from hand basins and canteen areas. Greywater from hand basins either provided to flush closets or collected separately for the treatment. As canteen or kitchen running mostly day and night times; produces more greywater; collected and treated separately.

Public amenities

In public amenities like bus stops, railway stations and airports, greywater is mainly generated from cleaning, hand basins and canteen areas. Greywater from hand basins is either provided to flush toilets or collected separately for the treatment. As canteens or kitchens running mostly day and night produce more greywater, it is collected and treated separately.

SOLAR POWER

Solar panels are installed on the rooftops of residential, commercial and industrial buildings. The power thus generated is utilized to fulfill the load requirement. In some cases, the power thus generated can be directly utilized for the non-critical loads, without the need of battery backup; these are called *direct-coupled solar PV systems*. Further, *battery-coupled solar PV systems* involve battery backup to maintain the continuity of power supply for the periods when no solar is available. The maintenance and the replacement costs get involved with battery storage.

Instead of utilizing battery backup, system utilizing grid power for periods when solar is unavailable are known as *grid-connected solar PV systems*. In most cases, the residential customer utilizes the rooftop solar PV system with the grid connection. During the solar power production period, the power is consumed from the solar PV, and during the night, power can be purchased from the grid. There are also *hybrid solar PV systems* which rely on auxiliary sources of power (fossil fuel generators or the grid). In this system, the battery storage is utilized to avoid the short term fluctuations. This type of system is most commonly used for critical applications or in places where large variation in the sunlight occurs throughout the year.

TREATMENT

Waste stabilization ponds (WSPs)

Waste or wastewater stabilization ponds (WSPs) are large, man-made water bodies in which blackwater, greywater or fecal sludge are treated by natural occurring processes and the influence of solar light, wind, micro-organisms and algae. In a typical WSP, three sequential ponds are involved: (a) the anaerobic pond (AP), which receives the influent and is in an anaerobic condition; (b) the facultative pond (FP) that receives the effluent from the AP, which is at least partially aerobic because oxygen is generated due to algae-mediated photosynthesis; and (c) maturation ponds, which are predominantly aerobic and complement the removal of organic material and improve the hygienic quality of the final effluent (Figure 6.1) (Dos Santos and Van Haandel, 2021). WSPs are relatively low-cost for operation and maintenance in comparison of other methods. They are very efficient in BOD and pathogen removal. They are ideal for tropical and subtropical climates, as they required relatively high intensity of the sunlight and high temperature for their efficiency (Mahapatra et al., 2022).

However, WSPs require a large surface area and need experts due to their complex design. The effluent still contains nutrients (e.g. N and P) and is therefore appropriate for reuse in agriculture, but not for direct recharge into surface water. Further, there is need for regularly cleaning of the scum that tends to float on the pond surface. WSPs have the problem of aquatic plants that occupy the whole pond and need to be regularly uprooted as they act as a barrier for the sunlight to penetrate the water column further down and might be an ideal habitat for mosquito breeding. The WSP's anaerobic ponds are designed to remove maximum solids or sludge from the wastewater, which leads to filling of the pond with sludge, so these ponds must be desludged in every 2–5 years, or upon reaching the accumulated solids up to one-third of the pond volume (Cavalcanti, 2003).

Constructed wetlands (CWs)

Constructed wetlands or artificial wetlands (Phytorid) are engineered systems designed to mimic the processes of natural wetlands, and involve plants for the

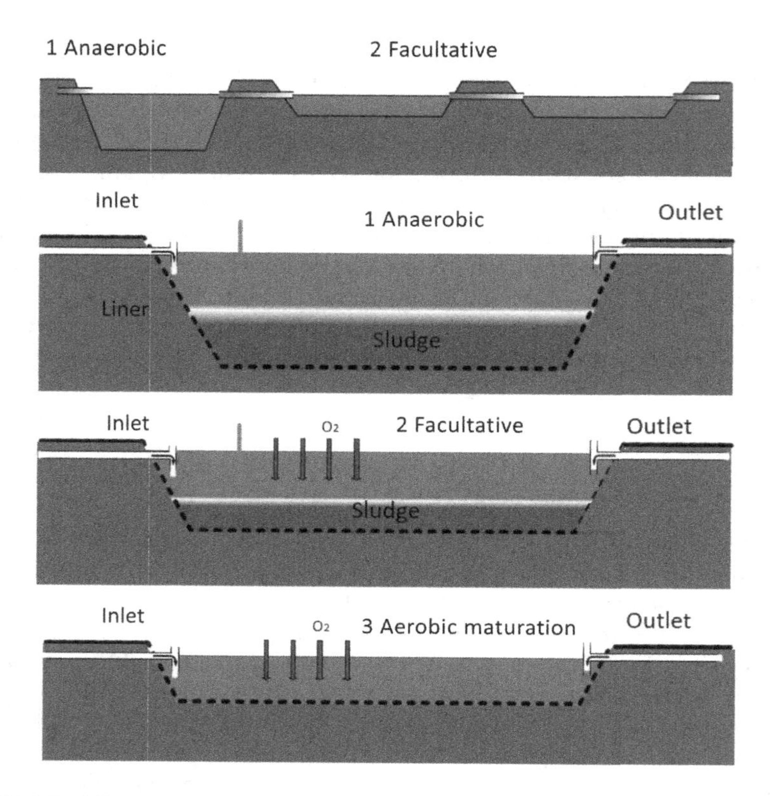

FIGURE 6.1 Wastewater stabilization pond

treatment of the wastewater in a process known as phytoremediation (Vymazal, 2019; Bakhshoodeh et al., 2020). It is a cost-effective and environmentally friendly approach that provides better treatment or polishing effects for wastewater that has been passed through primary treatment (septic tanks for blackwater) or secondary treatment; and are looking to achieve a higher quality effluent (Figure 6.2). It could be more suitable for small communities in urban, peri-urban and rural areas, where land is relatively cost-effective and easily available. This approach provides best results in warm climates, but they can be modified to adapt for limited freezing and periods of low biological activity.

There are various designs of constructed wetlands: (a) free-water surface CW; (b) horizontal subsurface-flow CW; and (c) vertical-flow CW (Vymazal, 2005, Kadlec and Wallace, 2008). A free-water surface CW designed on the basis of a natural wetland, marsh or swamp where naturally occurring processes take place. When wastewater gradually passes through the wetland, the particles tend to settle, pathogens are removed and the nutrients are utilized by micro-organisms and plants. In a horizontal subsurface-flow CW, a large basin of gravel and sand is prepared which have vegetation planted between gravel and sand-filled structures. When wastewater passes horizontally through the basin, the particles are filtered out by the filter

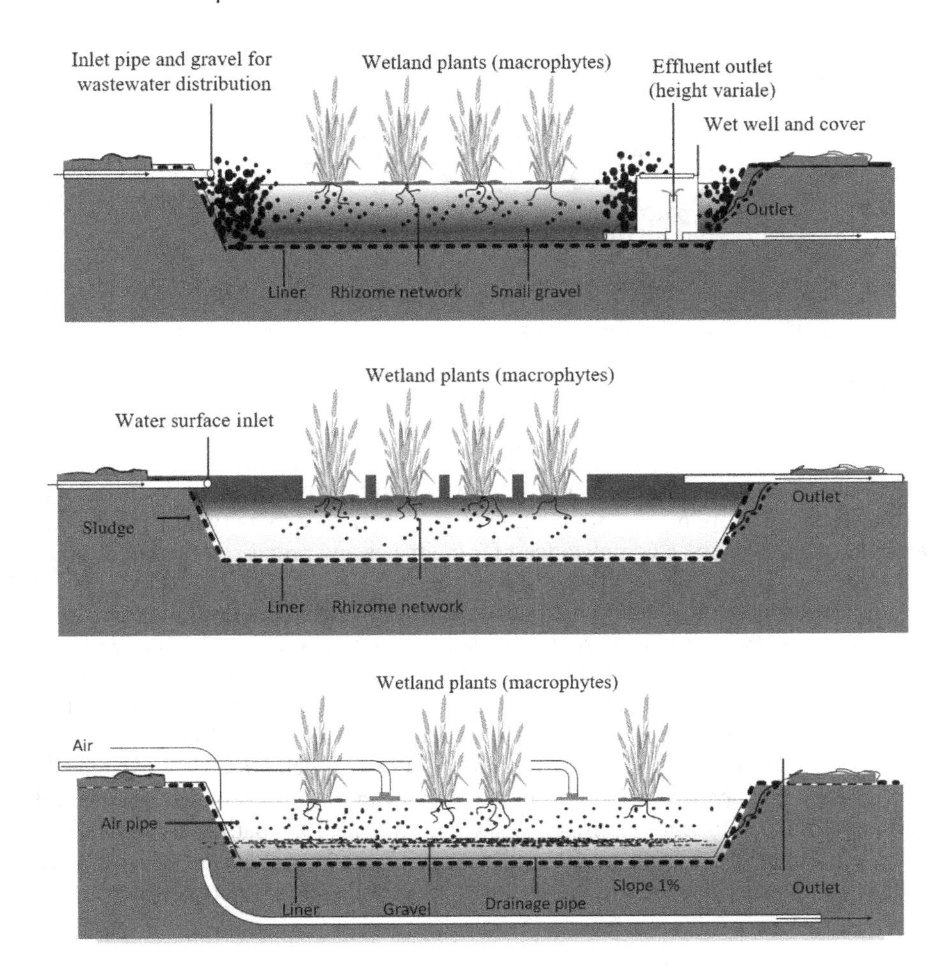

FIGURE 6.2 Constructed wetland

material and nutrients are degraded by the micro-organisms. In a vertical-flow CW, however, a filter bed that has vegetation like in free-flow and horizontal subsurface-flow CWs, but it is drained at the bottom; wastewater is supplied to the surface of the wetland through a mechanical dosing system, whereby it vertically passes down through the filter matrix to the bottom of the basin, collected by a drainage pipe. The important difference between vertical and horizontal wetlands is not simply the direction of the flow path, but rather the aerobic conditions.

Duckweed technology

Duckweed is small free-floating and robust-growth aquatic plants that have the great capability to purify wastewater with association of both aerobic and anaerobic micro-organisms (Sree and Appenroth, 2020). Duckweed-based wastewater treatment

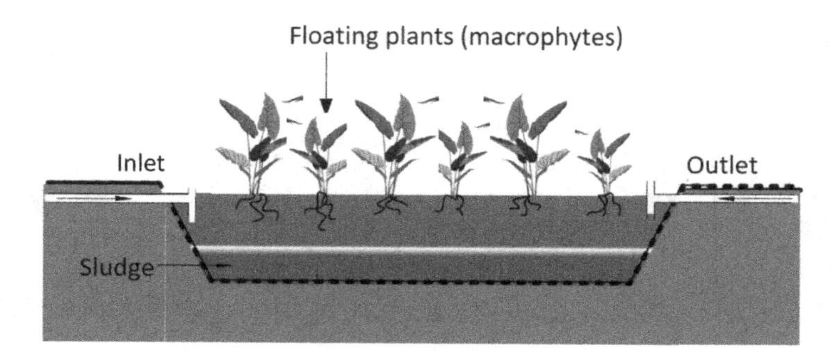

FIGURE 6.3 Duckweed technology

offers an effective treatment of the wastewater with direct economic returns and employment generation from pisciculture (Figure 6.3). Due to its richness in protein and vitamins, it can be a complete feed for the fishes and processed further as poultry feed.

Duckweed technology is a less expensive approach in construction, and further, its operation and maintenance are quite simple in comparison to other technologies. In comparison to other aquatic plants, duckweeds are less sensitive to low temperature, pH fluctuations, high nutrient levels, pests and diseases. Beside nutrient removal, this technology provides a highly nutritive vegetative material (35%–45% protein content) that can be utilized as feed for fish and poultry (Ziegler et al., 2017; Appenroth et al., 2017). Further, raw or processed duckweed can be used as mulch or a natural soil (organic) enrichment to nitrogen-rich environments. Low pathogen removal due to reduced light penetration causes duckweed to die off in cold weather conditions. Treatment capacity may be lost during high floods if the area is not protected.

Soil bio-technology (SBT)

Soil bio-technology (SBT) developed by Prof. HS Shankar of IIT, Mumbai (Shankar et al., 2002a, 2002b; Patnaik et al., 2003) involves three basic natural processes: photosynthesis, respiration and mineral weathering (Figure 6.4). SBT functions as a terrestrial system for the removal of suspended solids, organic and inorganic matter from the wastewater (Kadam et al., 2007; MoDWS, 2015; Parde et al., 2021). For that, soil micro-organisms are used along with macro-organism *Pheretima elongata*, a geophagus earthworm; both are interdependent and support each other.

SBT is also known as constructed soil filter (CSF). These terrestrial systems are built from are constructed from reinforced cement concrete (RCC), stone masonry or soil bunds. This technology involves a raw water tank, bioreactor containment, a treated water tank, piping and pumps. SBT can be operated in either decentralized or centralized mode batches and batch or continuous process modes. There are many advantages of the SBT process: (a) no requirement of mechanical aeration; (b) no foul smell and toxic waste generated; and (c) no sludge produced in this process – but it requires larger land area and higher capital cost (~1.3 times) for its establishment.

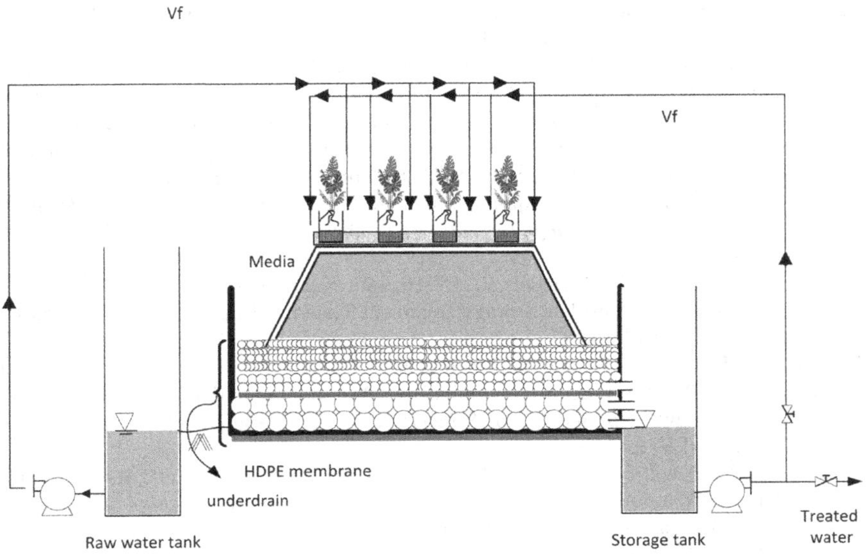

FIGURE 6.4 Soil bio-technology

REUSE OF TREATED WASTEWATER

Reuse of treated greywater has proven to be an established and sustainable approach to solve the water shortage problem around the world. Previously, treated wastewater was considered a disposable burden on treatment facilities or municipalities – but currently, recycled treated wastewater has proven to be a valuable resource. Treated greywater can be reused for toilet flushing, garden and landscape plant irrigation, floor and car washing, groundwater recharging and also in cooling towers in industry (Parjane and Sane, 2011; Karnapa, 2016; Edwin et al., 2014; Asano et al., 2007).

Friedler (2004) estimated that there is a reduction in freshwater demand of approximately 10%–20% if treated greywater is reused in toilet flushing. Reuse of greywater for toilet flushing and garden irrigation could be able to save up to 50% of total domestic water consumption (Maimon et al., 2010). When it is used for longer durations, there is a risk of increasing build-up of sodium and other micro-pollutants in soil which might affect plant growth (Maimon et al., 2010).

REFERENCES

Adegoke, A., and T. A. Stenstrom. 2019. "Cesspits and Soakpits." In *Global Water Pathogen Project*. Michigan State University, E. Lansing, MI: UNESCO. https://doi.org/10.14321/waterpathogens.58.

Appenroth, Klaus-J., K. Sowjanya Sree, Volker Böhm, Simon Hammann, Walter Vetter, Matthias Leiterer, and Gerhard Jahreis. 2017. "Nutritional Value of Duckweeds (Lemnaceae) as Human Food." *Food Chemistry* 217: 266–73. https://doi.org/10.1016/j.foodchem.2016.08.116.

Asano, T., F. Burton, and H. Leverenz. 2007. *Water Reuse: Issues, Technologies, and Applications*. New York: McGraw-Hill Education.

Bakhshoodeh, R., N. Alavi, C. Oldham, R. M. Santos, A. A. Babaei, J. Vymazal, and P. Paydary. 2020. "Constructed Wetlands for Landfill Leachate Treatment: A Review." *Ecological Engineering* 146: 105725. https://doi.org/10.1016/j.ecoleng.2020.105725.

Carr, R., and M. Strauss. 2001. "Excreta-Related Infections and the Role of Sanitation in the Control of Transmission. Water Quality" [Guidelines]. *Standards and Health*: 89–113.

Cavalcanti, P. F. F. 2003. *Integrated Application of the UASB Reactor and Ponds for Domestic Sewage Treatment in Tropical Regions*. Ph.D. thesis. Wageningen, The Netherlands: Wageningen University.

dos Santos, S. L., and A. van Haandel. 2021. "Transformation of Waste Stabilization Ponds: Reengineering of an Obsolete Sewage Treatment System." *Water* 13, no. 9: 1193. https://doi.org/10.3390/w13091193.

Edwin, G. A., P. Gopalsamy, and N. Muthu. 2014. "Characterization of Domestic Gray Water from Point Source to Determine the Potential for Urban Residential Reuse: A Short Review." *Applied Water Science* 4, no. 1: 39–49. https://doi.org/10.1007/s13201-013-0128-8.

Friedler, E. 2004. "Quality of Individual Domestic Greywater Streams and Its Implication for On-Site Treatment and Reuse Possibilities." *Environmental Technology* 25, no. 9: 997–1008. https://doi.org/10.1080/09593330.2004.9619393.

Kadam, A. M., P. D. Nemade, G. H. Oza, M. V. Sathyamoorthy, S. M. Dutta, and H. S. Shankar. 2007. "Wastewater Purification Using Soil Biotechnology System." In *National Conference at Govt*. Amravati, Maharashtra: College of Engineering.

Kadlec, R. H., and S. Wallace. 2008. *Treatment Wetlands*. 2nd Edition. Boca Raton: CRC Press.

Karnapa, A. 2016. "A Review on Gray Water Treatment and Reuse." *International Research Journal of Engineering & Technology* 3: 2665–8.

Loh, M., and P. Coghlan. 2003. *Domestic Water Use Study in Perth, Western Australia, 1998–2001*: 1–235. Perth: Water Corporation.

Mahapatra, Saswat, Kundan Samal, and Rajesh Roshan Dash. 2022. "Waste Stabilization Pond (WSP) for Wastewater Treatment: A Review on Factors, Modelling and Cost Analysis." *Journal of Environmental Management* 308: 114668. https://doi.org/10.1016/j.jenvman.2022.114668.

Maimon, Adi, Alon Tal, Eran Friedler, and Amit Gross. 2010. "Safe On-Site Reuse of Gray Water for Irrigation – A Critical Review of Current Guidelines." *Environmental Science & Technology* 44, no. 9: 3213–20. https://doi.org/10.1021/es902646g.

MoDWS. 2015. *Technical Options for Solid and Liquid Waste Management in Rural Areas*. New Delhi: Ministry of Drinking Water & Sanitation.

Parde, D., A. Patwa, A. Shukla, R. Vijay, D. J. Killedar, and R. Kumar. 2021. "A Review of Constructed Wetland on Type, Treatment and Technology of Wastewater." *Environmental Technology & Innovation* 21: 101261. https://doi.org/10.1016/j.eti.2020.101261.

Parjane, S. B., and M. G. Sane. 2011. "Performance of Gray Water Treatment Plant by Economical Way for Indian Rural Development." *International Journal of Chemical Technology Research* 3, no. 4: 1808–85.

Patnaik, B. R., U. S. Bhawalkar, A. Kadam, and H. S. Shankar. 2003. "Soil Biotechnology for Waste Water Treatment and Utilization." In *13th ASPAC 2003, International Water Works Association Conference*: 13–18 October. Quezon City, Philippines.

Poyyamoli, G., G. A. Edwin, and N. Muthu. 2013. "Constructed Wetlands for the Treatment of Domestic Grey Water: An Instrument of the Green Economy to Realize the Millennium Development Goals." In *The Economy of Green Cities: A World Compendium on the Green Urban Economy*: 313–21. Springer Dordrecht.

Shankar, H. S., B. R. Patnaik, and U. S. Bhawalkar. 2002a. "Process for Treatment of Organic Residues." India Patent Application MUM/384/26. April.

Shankar, H. S., B. R. Patnaik, and U. S. Bhawalkar. 2002b. "Process for Treatment of Waste Water." India Patent Application MUM/383/26. April.

Sree, K. S., and K. J. Appenroth. 2020. "Worldwide Genetic Resources of Duckweed: Stock Collections." In *The Duckweed Genomes*: 39–46. Springer.

Vymazal, J. 2005. "Horizontal Sub-surface Flow and Hybrid Constructed Wetlands Systems for Wastewater Treatment." *Ecological Engineering* 25, no. 5: 478–90. https://doi.org/10.1016/j.ecoleng.2005.07.010.

Vymazal, J. 2019. "Is Removal of Organics and Suspended Solids in Horizontal Sub-surface Flow Constructed Wetlands Sustainable for Twenty and More Years?" *Chemical Engineering Journal* 378: 122117. https://doi.org/10.1016/j.cej.2019.122117.

Ziegler, P., K. S. Sowjanya Sree, and K. J. Appenroth. 2017. "The Uses of Duckweed in Relation to Water Remediation." *Desalination & Water Treatment* 63: 327–42. https://doi.org/10.5004/dwt.2017.0479.

7 SPWR for blackwater

BLACKWATER RECYCLING

Domestic or municipal wastewater comprises the total water generated from houses, apartments, societies and shops, and restaurants or hotels, and it contains mainly organic load with limited presence of heavy metals and chemicals. It is categorized in two groups according to origin: (a) greywater; and (b) blackwater.

Blackwater includes the wastewater from toilets and lavatories; it can comprise feces, urine, water and toilet paper from flush toilets. As compared to greywater, it contains higher likely concentration of pathogens. Blackwater is further categorized into brown water (toilet wastewater without urine) and yellow water (urine with or without flush water).

COMPOSITION AND CHARACTERISTICS OF BLACKWATER

Blackwater contains half the load of organic material in domestic wastewater, the major fraction of the nutrients being nitrogen and phosphorus (Otterpohl et al., 1999; Kujawa-Roeleveld and Zeeman, 2006; Zuo et al., 2021), and can be collected with a small amount of water (1 liter per flush) using, for example, vacuum toilets. Blackwater also contains most of the pathogens, hormones and pharmaceutical residues of wastewater. The volume of blackwater depends on the type of toilet and amount of water needed to flush (Table 7.1).

COLLECTION OF BLACK WASTEWATER

INDIVIDUAL RESIDENTIAL AREAS

In individual houses (including in industry complexes), blackwater can be collected from toilets through separate pipes to in-house septic tanks or soak pits, or it can be directly conveyed to sewerage systems.

SHARED RESIDENTIAL AREAS

In apartments buildings and hostels, raw blackwater can be collected from toilets through shared drainage pipes to septic tanks or soak pits, or it can be directly conveyed to sewerage system.

COMMERCIAL AREAS

In commercial areas, blackwater is generated toilets but its composition might be different as compared to residential areas; there is less brown water from flushes and

DOI: 10.1201/9781003407690-7

TABLE 7.1
Blackwater characteristics

Parameter	Vacuum toilet 0.7–1.0 liters/flush (Wendland et al., 2007)	Flushing toilet 4–6 liters/flush (Otterpohl et al., 2004)
Chemical Oxygen Demand (COD) (mg/l)	10.76	1.827
Total Organic Content (TOC) (mg/l)	3.3	
Total Nitrogen (N_{total}) (mg/l)	1.54	274
Total Phosphorous (P_{total}) (mg/l)	254	46
Potassium ($K(K_2O)$) (mg/l)		85
Fecal coliforms (*E. coli*) Colony Forming Unit (CFU/100 ml)	2×10^8	

more yellow water from urinals. They are subcategorized into: (a) commercial areas like offices, institutions and shopping complexes, etc.; and (b) commercial areas like food shops, restaurants, hotels, etc.

Public amenities

In public amenities like bus stops, railway stations and airports, blackwater is generated in toilets but its composition might be different as compared to residential and commercial areas; there is less brown water from flushes and more yellow water from urinals.

SOLAR POWER

Solar panels can be installed on rooftops of residential, commercial or industrial buildings. The power thus generated is utilized to fulfill the load requirement. In some cases, the power thus generated can be directly utilized for the non-critical loads, without the need of battery backup, called *direct-coupled solar PV systems*. Further, *battery-coupled solar PV systems* involve a battery backup to maintain the continuity of power supply for the periods when no solar is available. Maintenance and replacement costs are involved with the battery storage.

Instead of utilizing the battery backup, *grid-connected solar PV systems* utilize the grid power for periods when solar is unavailable. In most cases, a residential customer utilizes a rooftop solar PV system with the grid connection. During the solar power production period, the power is consumed from the solar PV, and during night time, the power can be purchase from the grid. There are *hybrid solar PV systems* which rely on auxiliary sources of power (fossil fuel generators or the grid). In this system, battery storage is utilized to avoid the short-term fluctuations. This type of system is most commonly used for the critical applications or for the places where large variation in the sunlight occurs throughout the year.

TREATMENT

On-site systems

Septic tanks

Septic tanks are the underground and water sealed tanks pre-casted; those receive and partially treat the raw domestic blackwater (Bounds, 1997). Heavy solids sink to the bottom of the tank, while greases and lighter solids tend to float to the top; from there, the wastewater is either conveyed to the next tank or pit (in case of two-pit system) or sent to the sewer network for further treatment and is then discharged to soak pits or leach pits (Adhikari and Lohani, 2019).

Soak pits or leach pits

Soak pits or leach pits are the simplest method for the on-site partial treatment of the pre-settled blackwater effluent from septic tanks or centralized treatment facilities, which is then discharged to the underground perforated pit from which it can infiltrate to the surrounding soil. As the blackwater percolates from the pit to the soil, small particles are filtered out through the soil matrix, where microbial activity degrades the organic matter in the blackwater (Mondal, et al., 2014). These soak pits have proven to be ultimate solution for isolated areas and are best suited for soil with good absorptive properties but not appropriate for clay or hard-packed or rocky soil.

They are cost-effective and do not require skilled labor to construct. All of the wastewater from the septic tanks (after settling of maximum sludge) percolates into the soil matrix, leading to the groundwater recharging. These leach pits have the risk of groundwater contamination mainly in areas having high groundwater table conditions (Liu et al., 2008; Olatunji and Oladepo, 2013: Sajeesh and Pulikkal, 2022). To enhance the working life of soak pits or leach pits, a Nahani trap, floor trap or screening medium could be installed to separate out heavy solids (Uppala and Dey, 2021).

Centralized systems: STPs

Upflow anaerobic sludge blanket (UASB) reactor

UASB is a fast suspended growth treatment method in which pre-treated raw wastewater is supplied into the reactor from the bottom that it is uniformly distributed. In this approach, anaerobic bacteria (in flocs) are likely to suspend and maintain a blanket of anaerobic sludge at moderate flow rates, and the upward flow of wastewater helps to suspend the blanket of anaerobic sludge (Bal and Dhagat, 2001; Liu et al., 2003; Latif et al., 2011). Upon passing from the sludge blanket, particulate matter is retained and digested, while the breakdown of organic matter leads to production of gas and new sludge in relatively small amounts. As the gas bubbles rise upwards, they help in the mixing of the substrate with anaerobic biomass (Figure 7.1).

A UASB reactor is fitted with a gas collector dome and a secondary purification device constructing a gas/liquid/solid (GLS) phase that separates biogas, liquid fractions and sludge. The purified wastewater is collected in the troughs at the top of the reactor and removed. The biogas contains about 75% methane, which is collected in the dome and extracted for further use as fuel or flared (Schmidt and Ahring,

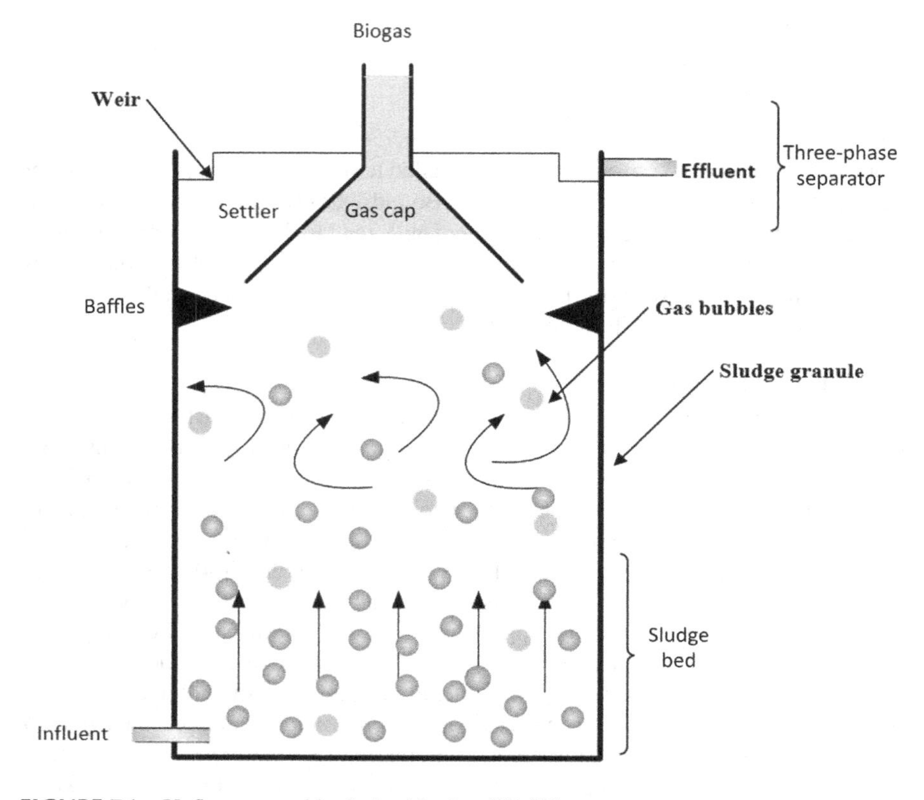

FIGURE 7.1 Upflow anaerobic sludge blanket (UASB) reactor

1993; Kaviyarasan, 2014). Due to small amounts of gas present in the settling zone, solid particles or sludge tends to fall back into the reactor. The main advantage of UASB reactor is that it has few moving parts and it does not require mechanical mixing as in other built or advanced treatment systems. If the influent's gravity distribution is possible, the UASB reactor may only need a pump to periodically remove excess sludge from the reactor to transfer it to the drying bed. Many UASB reactor plants are only able to remove 60%–70% of the BOD and TSS, further suitable post treatments methods like wastewater stabilization ponds and duckweed ponds. To eliminate this remaining 20–40% BOD and TSS, a one-day HRT pond is sufficient, which requires a very large area to build. UASB has proven to be very promising in rural areas where treated wastewater is commonly used for irrigation.

Extended aeration (EA)

The extended aeration process is a modified ASP which involves biological treatment for the removal of BOD and TSS under aerobic conditions. To provide the aerobic conditions, oxygen is supplied through surface or diffuse aeration to facilitate aerobic biological processes (Sotirakou et al., 1999; Zeinaddine et al., 2013). In this approach, raw wastewater is primarily screened and degritted to eliminate the large

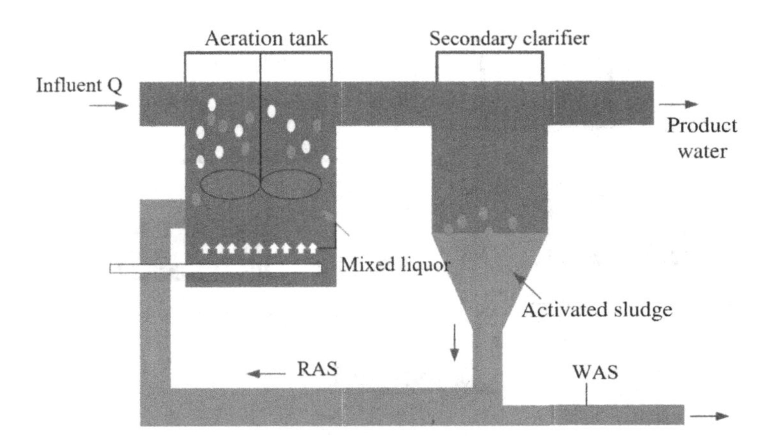

FIGURE 7.2 Extended aeration (EA) process

suspended, settleable or floating solids which could interfere with or damage the equipment used in the downstream process. If required, there is a provision of equalization basins, through flow of the effluent up to peak flow rates. Then the wastewater reached in the aeration tank, where mixing and aeration takes place (Figure 7.2).

The mixed wastewater is conveyed to a sedimentation tank, then the purified wastewater flows over a weir and and into a collection tank before being sent to the disinfection process. In this process, most of the flocculated materials (with colonies or micro-organism) settle down to the bottom and a part of it (return activated sludge [RAS]) is returned back to the influent wastewater at the start of the process (US EPA, 2000). The remaining material (waste activated sludge [WAS]) is collected for sludge treatment and disposal. The extended aeration process is considered suitable for such treatment plants, especially smaller treatment plants that contain low amounts of settleable solids in their raw wastewater, leading to a reduction in the number of unit processes followed in the treatment plants. The EA process is also very reliable in performance with specialization in handling of relatively high hydraulic shock loads and high organic content; plus, there is no waste of biomass.

Sequencing batch reactor (SBR)

Sequencing batch reactor (SBR) is a modified ASP in which all three processes – equalization process, aeration process and clarification process –take place in a single reactor tank. It involves a cycle of five phases: (a) fill; (b) react; (c) settle; (d) draw; and (e) idle. Sludge is collected during the draw (decant) phase of the cycle (Singh and Srivastava, 2011; Fernandes et al., 2013). The wastewater is supplied to the reactor tank, where all the required treatment processes is followed and purified wastewater is pumped out from the reactor tank (Figure 7.3).

Each phase of the SBR cycle is designated with a definite time period (depending on the aeration and mixing pattern). The additional duration of aeration time depends on the size of the plant and the amount/composition of inflowing wastewater, but is typically 60–90 minutes. SBR systems are suitable for low flow rates of 0.1 MLD,

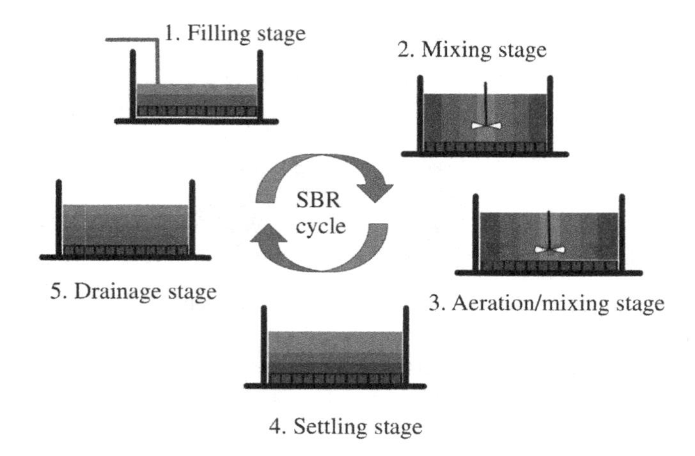

FIGURE 7.3 Sequencing batch reactor (SBR)

and provide more operational flexibility and minimal space requirements. Due to being fully automated, SBRs do not require much manpower for maintenance.

REUSE OR SALE

Treating wastewater or reclaiming it for reuse has proven to be an efficient and sustainable approach for solving the water scarcity crisis worldwide. While before it was believed to be a burden to discard, nowadays treated or reclaimed wastewater can be an important resource which can be used in irrigating agricultural crops, fruit gardens and plantations, golf courses and parks. Beside these uses, treated or reclaimed wastewater can be also applied in groundwater recharging and conservation or expanding wetlands or riparian habitats, plus in residential areas for street cleaning, gardening, toilet flushing and other related uses.

AGRICULTURE

Being an agricultural economy, India has an appealing opportunity to use treated or reclaimed wastewater (Jaramillo and Restrepo, 2017), but this should be avoided in irrigation of edible crops such as cereals or millets, etc. (Anjana and Iqbal, 2007; Becerra-Castro et al., 2015). The reuse of treated or reclaimed wastewater for most agricultural crops is an effective approach but it should be encouraged locally where it is generated and its inter-basin sharing of reuse benefits should be avoided due to water rights and other problems. In comparison to urban areas, rural areas are more relevant for agricultural reuse – and for this, local wastewater first treated in waste stabilization ponds and then in maturation ponds provides for best results. Before familiarizing yourself with agricultural reuse practices, there following things should be considered.

1 Year-round investigation of treated water reuse in agriculture should take place through using rotational crop systems and planned treatment facilities as such that there is minimum wastage of treated sewage during summer.
2 The manual direct handling of treated wastewater should be discouraged, and for the transport of such wastewater, field channels are more suitable than more complicated drip irrigation systems, etc.
3 The discharge standards and permissible limits based on important parameters should be followed carefully before the treated wastewater is used in irrigation.

Fish culture

The reuse of treated or reclaimed wastewater for raising fish, or pisciculture, is another good and acceptable approach. The best example of this practice is Kolkata city's wetlands that are unique and natural wastewater treatment systems which also provide fish and vegetables for the city (Bunting et al., 2010). Fish ponds are efficient in treating the municipal wastewater, but these ponds cannot be considered as standalone treatment systems; it is better for the wastewater to be first treated with stabilization ponds and maturation ponds, then fish be raised in the treated or reclaimed wastewater. Further, there is strict and regular monitoring by the concerned authority to reuse treated or reclaimed diluted sewage for the pisciculture along the lines of the ongoing East Kolkata wetlands.

Discharge into non-perennial/dry rivers

Treated or reclaimed wastewater could be discharged in to non-perennial stream channels to make regular flow for rejuvenating the dry rivers, where it percolates into the aquifer zone and later is pulled out by infiltration wells or galleries. Due to additional purification by the soil and further dilutions, aquifer water provides a regular flow to rejuvenate the dry rivers. For such purposes, the following guiding principles for the ULBs should be observed.

1 Monitoring the quality of the raw water from the infiltration wells to check the unexpected rise in contaminants.
2 Intercepting and treating the wastewater outfalls through well-defined sewerage networks and the STPs.

REFERENCES

Adhikari, J. R., and S. P. Lohani. 2019. "Design, Installation, Operation and Experimentation of Septic Tank–UASB Wastewater Treatment System." *Renewable Energy* 143: 1406–15. https://doi.org/10.1016/j.renene.2019.04.059.

Anjana, S. U., and M. Iqbal. 2007. "Nitrate Accumulation in Plants, Factors Affecting the Process, and Human Health Implications. A Review." *Agronomy for Sustainable Development* 27, no. 1: 45–57. https://doi.org/10.1051/agro:2006021.

Bal, A. S., and N. N. Dhagat. 2001. "Upflow Anaerobic Sludge Blanket Reactor a Review." *Indian Journal of Environmental Health* 43, no. 2: 1–82.

Becerra-Castro, Cristina, Ana Rita Lopes, Ivone Vaz-Moreira, Elisabete F. Silva, Célia M. Manaia, and Olga C. Nunes. 2015. "Wastewater Reuse in Irrigation: A Microbiological Perspective on Implications in Soil Fertility and Human and Environmental Health." *Environment International* 75: 117–35. https://doi.org/10.1016/j.envint.2014.11.001.

Bounds, T. R. 1997. "Design and Performance of Septic Tanks." *ASTM Special Technical Publication* 1324: 217–34.

Bunting, S. W., J. Pretty, and P. Edwards. 2010. "Wastewater-Fed Aquaculture in the East Kolkata Wetlands, India: Anachronism or Archetype for Resilient Ecocultures?" *Reviews in Aquaculture* 2, no. 3: 138–53. https://doi.org/10.1111/j.1753-5131.2010.01031.x.

Fernandes, Heloísa, Mariele K. Jungles, Heike Hoffmann, Regina V. Antonio, and Rejane H. R. Costa. 2013. "Full-Scale Sequencing Batch Reactor (SBR) for Domestic Wastewater: Performance and Diversity of Microbial Communities." *Bioresource Technology* 132: 262–8. https://doi.org/10.1016/j.biortech.2013.01.027.

Jaramillo, M. F., and I. Restrepo. 2017. "Wastewater Reuse in Agriculture: A Review about Its Limitations and Benefits." *Sustainability* 9, no. 10: 1734. https://doi.org/10.3390/su9101734.

Kaviyarasan, K. 2014. "Application of UASB Reactor in Industrial Wastewater Treatment–A Review." *International Journal of Scientific & Engineering Research* 5, no. 1: 584.

Kujawa-Roeleveld, K., and G. Zeeman. 2006. "Anaerobic Treatment in Decentralised and Source-Separation-Based Sanitation Concepts." *Reviews in Environmental Science & Bio/Technology* 5, no. 1: 115–39. https://doi.org/10.1007/s11157-005-5789-9.

Latif, Muhammad Asif, Rumana Ghufran, Zularisam Abdul Wahid, and Anwar Ahmad. 2011. "Integrated Application of Upflow Anaerobic Sludge Blanket Reactor for the Treatment of Wastewaters." *Water Research* 45, no. 16: 4683–99. https://doi.org/10.1016/j.watres.2011.05.049.

Liu, Y., Hai-Lou Xu, Shu-Fang Yang, and Joo-Hwa Tay. 2003. "Mechanisms and Models for Anaerobic Granulation in Upflow Anaerobic Sludge Blanket Reactor." *Water Research* 37, no. 3: 661–73. https://doi.org/10.1016/s0043-1354(02)00351-2.

Liu, Z. B., H. K. Yan, and Z. J. Wang. 2008. "Research on Ground Water Pollution by Leacheate of Waste Dump of Open Pit Coal Mine." *Journal of Coal Science & Engineering (China)* 14, no. 1: 114–8. https://doi.org/10.1007/s12404-008-0023-3.

Mondal, P., A. Nandan, N. A. Siddiqui, and B. P. Yadav. 2014. "Impact of Soak Pit on Groundwater Table. Environ." *Pollution Control Journal* 18: 12–7.

Olatunji, J. J., and K. T. Oladepo. 2013. "Microbiological Quality of Water Collected from Unlined Wells Located near Septic-Tank Soak Away and Pit Latrines in Ife North Local Government Area of Osun State, Nigeria." *Transnational Journal of Science & Technology* 3, no. 10: 8–19.

Otterpohl, R., A. Albold, and M. Oldenburg. 1999. "Source Control in Urban Sanitation and Waste Management: Ten Systems with Reuse of Resources." *Water Science & Technology* 39, no. 5: 153–60. https://doi.org/10.2166/wst.1999.0234.

Otterpohl, R., U. Braun, and M. Oldenburg. 2004. "Innovative Technologies for Decentralised Water-, Wastewater and Biowaste Management in Urban and Peri-urban Areas." *Water Science & Technology* 48, no. 11–12: 23–32. https://doi.org/10.2166/wst.2004.0795.

Sajeesh, A. K., and A. K. Pulikkal. 2022. "Water Quality Assessment of Open Wells in Malappuram District, Kerala." *AQUA — Water Infrastructure, Ecosystems and Society* 71, no. 12: 1325–31.

Schmidt, J. E., and B. K. Ahring. 1993. "Effects of Hydrogen and Formate on the Degradation of Propionate and Butyrate in Thermophilic Granules from an Upflow Anaerobic Sludge Blanket Reactor." *Applied & Environmental Microbiology* 59, no. 8: 2546–51. https://doi.org/10.1128/aem.59.8.2546-2551.1993.

Singh, M., and R. K. Srivastava. 2011. "Sequencing Batch Reactor Technology for Biological Wastewater Treatment: A Review." *Asia-Pacific Journal of Chemical Engineering* 6, no. 1: 3–13. https://doi.org/10.1002/apj.490.

Sotirakou, E., G. Kladitis, N. Diamantis, and H. Grigoropoulou. 1999. "Ammonia and Phosphorus Removal in Municipal Wastewater Treatment Plant with Extended Aeration. Global Nest." *International Journal* 1, no. 1: 47–53.

Uppala, P., and S. Dey. 2021. "Design of Potential Rainwater Harvesting Structures for Environmental Adoption Measures in India." *Polytechnica* 4, no. 2: 59–80. https://doi.org/10.1007/s41050-021-00035-9.

US. EPA. 2000. *Methods for Measuring the Toxicity and Bioaccumulation of Sediment-Associated Contaminants with Freshwater Invertebrates*. 2nd ed., EPA 600/R-99/064. Washington, DC: US EPA.

Wendland, C., S. Deegener, J. Behrendt, P. Toshev, and R. Otterpohl. 2007. "Anaerobic Digestion of Blackwater from Vacuum Toilets and Kitchen Refuse in a Continuous Stirred Tank Reactor (CSTR)." *Water Science & Technology* 55, no. 7: 187–94. https://doi.org/10.2166/wst.2007.144.

Zeinaddine, H. R., A. Ebrahimi, V. Alipour, and L. Rezaei. 2013. "Removal of Nitrogen and Phosphorous from Wastewater of Seafood Market by Intermittent Cycle Extended Aeration System (ICEAS)." *Journal of Health Sciences & Surveillance System* 1, no. 2: 89–93.

Zuo, S., X. Zhou, Z. Li, X. Wang, and L. Yu. 2021. "Investigation on Recycling Dry Toilet Generated Blackwater by Anaerobic Digestion: From Energy Recovery to Sanitation." *Sustainability* 13, no. 8: 4090. https://doi.org/10.3390/su13084090.

8 SPWR for industrial wastewater

INDUSTRIAL WASTEWATER

Industrial wastewater has different constituents and variable characteristics which change depending on the type of industry and processes used in that industry. It could contain high amounts of organic load, dyes, heavy metals, volatile compounds and synthetic chemicals that prove to be toxic to very toxic in nature (Table 8.1). According to the industry where it was generated, it can be called distillery wastewater, tannery wastewater, pulp wastewater, textile wastewater or petrochemical wastewater. There are separate standards for the permissible limits of pollutants presented in wastewater from the different industries.

CHARACTERISTICS AND COMPOSITION OF INDUSTRIAL WASTEWATER

The composition and nature of industrial wastewater changes from time to time, and further, it does not have a uniform or continuous discharge rate; thus, the wastewater might need equalization process prior to the treatment. In the equalization step, the wastewater needs to be held for a specific period in a continuously mixing tank that provides a fairly uniform wastewater flow. If the wastewater is of a very acidic or alkali nature (mainly if it is acidic), the wastewater needs to undergo a neutralization process in the neutralization tank. It is also possible to perform the neutralization process in the equalization tank, if conditions allow.

If the industrial wastewater is combined with municipal wastewater – treated together or released into a stream – then proportioning might be used as a unit process for the wastewater. Proportioning includes the management of industrial wastewater's release into the receiving sewer or stream, in a defined proportion to the flow of municipal wastewater or stream. There are two benefits of such proportioning: (a) shielding the treatment equipment from shock loads; and (b) improving the hygienic quality of the treated wastewater.

INDUSTRIAL WATER RECYCLING

COLLECTION

Solar power

Solar panels can be installed on the rooftops of residential, commercial and industrial buildings. The power thus generated is utilized to fulfill the load requirements. In some cases, the power thus generated can be directly utilized for non-critical

DOI: 10.1201/9781003407690-8

TABLE 8.1
The important characteristics of different types of industrial wastewater and their treatment methods

Industrial wastewater	Major characteristics	Major treatment and disposal methods
Textile	Highly alkaline, colored, COD, temperature, high suspended solids	Neutralization, chemical precipitation, biological treatment, aeration, and/or trickling filtration
Tannery	High total solids, hardness, salt sulfides, chromium, pH, precipitated lime, and BOD5	Equalization, sedimentation and biological treatment
Dairy	High in dissolved organic matter, mainly protein, fat and lactose	Acidification, flotation biological treatment, aeration trickling filtration, activated sludge
Meat and poultry	High in dissolved and suspended organic matter, blood, other proteins and fats	Screening, setting and/or flotation, trickling filtration
Distillery	High in dissolved organic solids containing nitrogen and fermented starches or their products	Recovery, concentration by centrifugation and evaporation, trickling filtration, use in feeds, digestion of slops
Pulp and paper	High or low pH, color, high suspended, colloidal and dissolved solids, inorganic filters	Settling lagooning, biological treatment, aeration, recovery of byproducts using flotation
Sugar	Variable pH, should organic matter with relatively high BOD5 of carbonaceous nature	Neutralization, calculation, chemical treatment, some selected aerobic oxidation
Pharmaceuticals	High in suspended and dissolved organic matter	Activated sludge
Petrochemicals	High COD, TDS, metals, COD/ BOD5 ratio	Recovery and reuse, equalization and neutralization, chemical coagulation, settling or flotation, biological oxidation
Pesticides	High organic matter, benzene ring structure, toxic to bacteria and fish, acidic	Activated carbon adsorption, alkaline chlorination

loads, without the need of battery backup, called *direct-coupled solar PV systems*. Further, *battery-coupled solar PV systems* involve a battery backup to maintain the continuity of power supply for the periods when no solar is available. Maintenance and replacement costs become involved with battery storage.

Instead of utilizing battery backup, systems utilizing grid power for periods when solar is unavailable are known as *grid-connected solar PV systems*. In most cases, the residential customer utilizes a rooftop solar PV system with the grid connection. During the solar power production period, the power is consumed from the solar PV, and during the night, power can be purchase from the grid. There are also *hybrid solar PV systems* which rely on auxiliary sources of power (fossil fuel generators or the grid). In this system, battery storage is utilized to avoid short-term fluctuations. This type of system is most commonly used for critical applications or for places where large variations in amounts of sunlight occur throughout the year.

TREATMENT

Activated sludge process (ASP)

The activated sludge process (ASP) is a technology that was devised a century ago to treat the sewage and industrial wastewaters (shown in Figure 8.1). There could be different variations in the designs, but all ASPs have three main elements: (a) an aeration tank to perform the biological treatment; (b) a settling tank to settle out the solids (activated sludge AS) and provide treated wastewater; and (c) an RAS mechanism that convey some part of settled AS from the clarifier to the influent wastewater in the aeration tank (Eckenfelder, 1998; Santos and Judd, 2010). Due to biological (and sometimes chemical) processes, the proportion of influent which is biodegradable is significantly reduced in the aeration tank.

In the ASP, oxygen or air is blown into the untreated raw wastewater in an aeration tank. The micro-organism starts the breakdown of solid particles (mainly biodegradable matter) into smaller particles. Then it is conveyed in to settling tank where the activated sludge accelerates the degradation process. In this process, live bacterial colonies tend to go downwards to the bottom of the settling tank, which is returned back to the digester; meanwhile, dead bacterial colonies float to the surface of the tank and the treated clean water is supplied to another tank for further treatment or discharged into the drains.

In this method of wastewater treatment, to understand the operation of entire ASP, it is quite essential to understand activated sludge and its mechanism. The sludge particles having live sludge bacterial colonies are colonies actively swarming in to the wastewater to degrade the biodegradable matter; the sludge is termed as activated sludge (Hreiz et al., 2015).

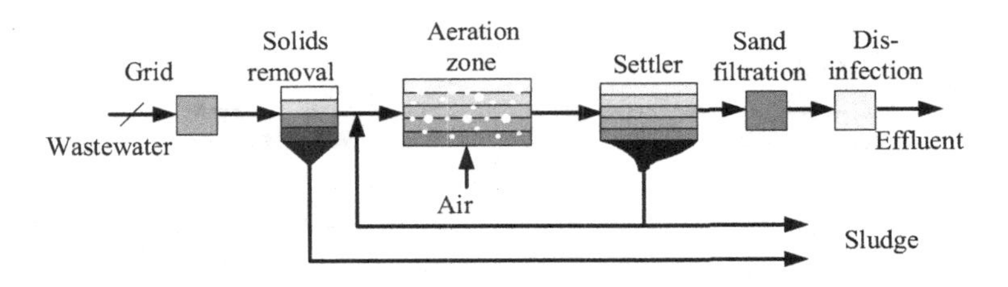

FIGURE 8.1 Activated sludge process

Membrane bioreactors (MBRs)

Membrane bioreactors (MBR) use a perm-selective or semi-permeable membrane, along with a bioreactor (Santos and Judd, 2010). This method involves microfiltration or ultrafiltration through a membrane process and biological degradation through a suspended growth bioreactor (Shown in Figure 8.2). Due to the removal of secondary clarifiers and tertiary filtration processes, they are gaining interest and are also used for large-scale operations for the treatment of municipal and industrial wastewater (Le-Clech, 2010; Park et al., 2015). MBR technology could help upgrade old sewage treatment plants and reduce their plant footprints.

In a membrane bioreactor, the membranes are sunken in an aeration tank, which have the 0.035–0.4 μ pores that are considered between microfiltration and ultrafiltration. These types of membranes permit a high quality wastewater, which eliminates the need for sedimentation and filtration processes that are widely implicated in the wastewater treatment. Due to elimination of the sedimentation process, a more homogenous liquor concentration is obtained for the biological treatment process, resulting in the reduction of the process tanks that have traditionally been required for the consequent processes. Due to this, existing plants can be advanced without the construction of new process tanks. For facilitating the optimal aeration and flushing around membranes, the wastewater is mainly maintained in the range of 1.0–1.2% solids, or four times that of conventional equipment.

Advanced oxidation processes (AOPs)

Advanced oxidation processes (AOPs) are an advanced and encouraging approach for treating wastewater pollutants, especially organic pollutants in industrial wastewater (Figure 8.3). For the effective degradation of organic pollutants, this approach relies on in situ production of strong oxidants such as hydroxyl radicals and sulfate radicals (Poyatos et al., 2010; Wang and Xu, 2012). Beside the utilization of strong oxidants, AOPs technology also uses ozone and UV radiation for better treatment of wastewater. Industrial wastewater is known to have numerous harmful chemicals that have been proven to have very toxic effects on aquatic biota and human health. Therefore, there is a requirement of such technologies such as AOPs that are capable of removing toxic pollutants from wastewater, especially industrial wastewater.

There are many AOPs technologies such as photocatalysis, Fenton-like processes and ozonation (Wang et al., 2016; Ameta et al., 2018) that have been explored in wastewater treatment in order to provide an efficient and integrated approach for

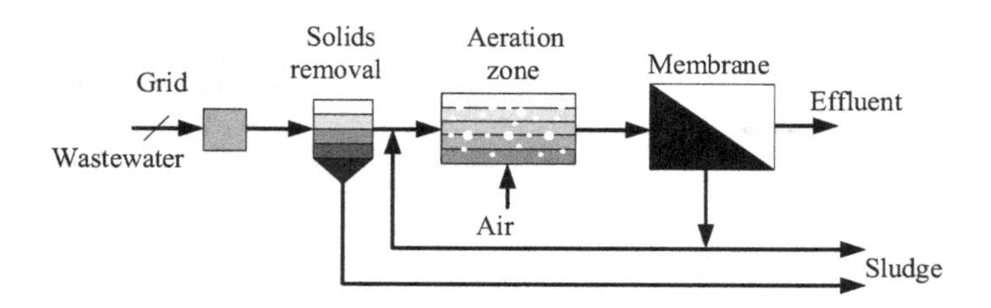

FIGURE 8.2 Membrane bioreactor (Santos and Judd, 2010)

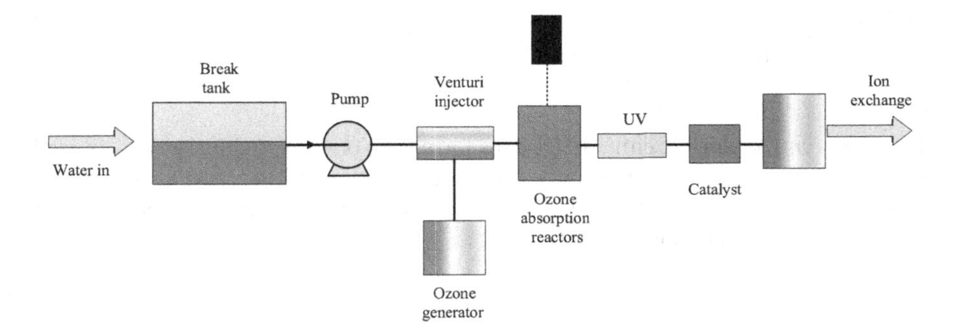

FIGURE 8.3 Advanced oxidation processes

industrial wastewater treatment that contains toluene, xylene, naphthalene, phenol, malonic acid and acetic acid.

Common effluent treatment plants (CETPs)

Common effluent treatment plants (CETPs) are the shared effluent treatment systems that primarily plan for the collective treatment of wastewater of an industrial complex that comprises many small industrial facilities (Vyas et al., 2011; Padalkar and Kumar, 2018). There are several problems running an individual effluent treatment plant (ETP) which small industrial facilities generally face such as lack of space, resources, investment costs and skilled manpower for operation and maintenance (Pathe et al., 2004). Through the use of CETPs, these problems are overcome by the collective treatment of effluents from large numbers of small industrial facilities at a single place, which helps in reducing land requirements and investment costs, and also provides efficient wastewater treatment and ease of operation and maintenance (Padalkar et al., 2016).

CETPs can be divided into two types: (a) homogenous CETPs that receive the same effluent from numerous similar industrial activities inside a industrial complex; and (b) heterogeneous CETPs that process different effluents from numerous industrial activities which are involved in the production of different products.

CETP can help in showing consistency in the treatment facility's ability to forecast the treatment values or ranges to which it will adapt in the future. When the pollutant levels in the effluent treated by the facility satisfy the established discharge criteria, CETP is said to be more reliable than an individual ETP, which frequently exceeds the standards. CETP has proven to be very useful in predicting the nature of a treatment facility in the future and also helps in setting a threshold value for a given facility. In comparison of individual ETPs, the cost of the addition of chemicals is reduced, leading to lower operating costs. It is also observed that continuous microbe seeding in the CETP demonstrates greater control over the wastewater quality.

Disinfection

Disinfection of black wastewater is accomplished by several methods (Liberti and Notarnicola, 1999), including: (a) chemical processes, such as ozonation and chlorination; (b) physical processes, such as UV radiation and microfiltration; and (c) biological processes, such as aerated lagoons).

Chemical

Chlorination Chlorine, either in gaseous form (Cl_2) or as hypochlorite salts, is used to disinfect wastewater. All types of chlorine react with water to create hypochlorous acid (HOCl), which quickly undergoes breakdown to create the hypochlorite ion according to the following reaction.

$$HOCl \leftrightarrow OCl^- + H^+$$

Chlorine is far more effective in killing enteric bacteria like *E. coli* than other bacterial species. Therefore, the comparative sensitivities of the various pathogen communities must be taken into account when using *E. coli* to evaluate disinfection efficiency. The ideal ratio of pH, chlorine concentration and contact duration, as well as the concentrations of ammonia and suspended particles, are all necessary for effective chlorine disinfection (Huang et al., 2011).

Ozonation Ozonation is used to disinfect surfaces through generating free radicals as oxidizing agents. In comparison to chlorination, ozonation is more effective in inhibiting the viruses and bacteria, but when conditions are not right, the effective bactericidal activity can have issues (Agustina et al., 2005). The key problem that significantly inhibits ozone's ability to disinfect water is its low solubility, and that any residual ozone produced quickly evaporates due to its reactive nature. It may be difficult to gauge the effectiveness of the disinfection procedure and may allow for potential microbial re-growth, which is another drawback of the lack of a persistent residual.

Physical

Ultraviolet radiation The physical process of disinfecting treated wastewater with UV radiation requires passing a film of the wastewater in close proximity to a UV source (lamp). The physical and chemical water quality features of the wastewater before treatment determine how effectively UV disinfection works. An improved effluent quality leads to a more effective UV disinfection procedure. The UV disinfection method has the advantage of being quick without increasing the toxicity of the effluent (Das, 2001).

REUSE OR SALE

BOILER FEED WATER

Treated wastewater could be used as boiler feed water, which might need pre-treatment before being used for cooling applications (Liu et al., 2008). The quality of boiler feed water primarily depends on the boiler pressures at which the steam is to be generated. Increasing the boiler pressure increases the purity of the water produced. If treated water has hardness more than 40 mg/l, mixing with soft water is required to achieve the desired hardness.

Process water

The treated or reclaimed wastewater also used in some industrial processes. These processes can be identified by screening such processes that do not need high-quality fresh water and can be achieved with low-quality treated or reclaimed wastewater. For this, there is need for multiple quality water supply specifications within the industrial setup.

Cooling water

The reuse of treated or reclaimed wastewater as cooling water is another suitable and sustainable approach in industrial applications. If treated water has more alkalinity, acid treatment is required before it can be used in the cooling towers. However, there is problem of increasing the sulfate content somewhat due to use of sulfuric acid (H_2SO_4) (CPHEEO, 1993).

REFERENCES

Agustina, T. E., H. M. Ang, and V. K. Vareek. 2005. "A Review of Synergistic Effect of Photocatalysis and Ozonation on Wastewater Treatment." *Journal of Photochemistry & Photobiology C* 6, no. 4: 264–73. https://doi.org/10.1016/j.jphotochemrev.2005.12.003.

Ameta, R., M. S. Solanki, S. Benjamin, and S. C. Ameta. 2018. "Photocatalysis." In *Advanced Oxidation Processes for Waste Water Treatment*: 135–75. Cambridge, MA: Academic Press.

CPHEEO. 1993. *Manual on Sewerage and Sewage Treatment*. New Delhi: Ministry of Urban Development.

Das, T. K. 2001. "Ultraviolet Disinfection Application to a Wastewater Treatment Plant." *Clean Products & Processes* 3, no. 2: 69–80. https://doi.org/10.1007/s100980100108.

Eckenfelder, W. 1998. *Activated Sludge: Process Design and Control*. Boca Raton: CRC Press.

Hreiz, R., M. A. Latifi, and N. Roche. 2015. "Optimal Design and Operation of Activated Sludge Processes: State-of-the-Art." *Chemical Engineering Journal* 281: 900–20. https://doi.org/10.1016/j.cej.2015.06.125.

Huang, Jing-Jing, Hong-Ying Hu, Fang Tang, Y. Li, Sun-Qin Lu, and Yun Lu. 2011. "Inactivation and Reactivation of Antibiotic-Resistant Bacteria by Chlorination in Secondary Effluents of a Municipal Wastewater Treatment Plant." *Water Research* 45, no. 9: 2775–81. https://doi.org/10.1016/j.watres.2011.02.026.

Le-Clech, Pierre. 2010. "Membrane Bioreactors and Their Uses in Wastewater Treatments." *Applied Microbiology & Biotechnology* 88, no. 6: 1253–60. https://doi.org/10.1007/s00253-010-2885-8.

Liberti, L., and M. Notarnicola. 1999. "Advanced Treatment and Disinfection for Municipal Wastewater Reuse in Agriculture." *Water Science & Technology* 40, no. 4–5: 235–45. https://doi.org/10.2166/wst.1999.0596.

Liu, H., C. Yang, W. Pu, and J. Zhang. 2008. "Removal of Nitrogen from Wastewater for Reusing to Boiler Feed-Water by an Anaerobic/Aerobic/Membrane Bioreactor." *Chemical Engineering Journal* 140, no. 1–3: 122–9. https://doi.org/10.1016/j.cej.2007.09.048.

Padalkar, A. V., and R. Kumar. 2018. "Common Effluent Treatment Plant (CETP): Reliability Analysis and Performance Evaluation." *Water Science & Engineering* 11, no. 3: 205–13. https://doi.org/10.1016/j.wse.2018.10.002.

Padalkar, A. V., K. Satinder, and K. Rakesh. 2016. "Performance Evaluation of Common Effluent Treatment Plant for Efficiency of Pollutant Removal and Relation to Design Adequacy." *Journal of Environmental Science & Engineering* 58, no. 1: 17–28.

Park, H. D., I. S. Chang, and K. J. Lee. 2015. *Principles of Membrane Bioreactors for Wastewater Treatment*. Boca Raton: CRC Press.

Pathe, P. P., M. Suresh Kumar, M. R. Kharwade, and S. N. Kaul. 2004. "Common Effluent Treatment Plant (CEPT) for Wastewater Management from a Cluster of Small Scale Tanneries." *Environmental Technology* 25, no. 5: 555–63. https://doi.org/10.1080/09593332608618562c.

Poyatos, J. M., M. M. Muñio, M. C. Almecija, J. C. Torres, E. Hontoria, and F. Osorio. 2010. "Advanced Oxidation Processes for Wastewater Treatment: State of the Art." *Water, Air, & Soil Pollution* 205, no. 1–4: 187–204. https://doi.org/10.1007/s11270-009-0065-1.

Santos, Ana, and Simon Judd. 2010. "The Fate of Metals in Wastewater Treated by the Activated Sludge Process and Membrane Bioreactors: a Brief Review." *Journal of Environmental Monitoring* 12, no. 1: 110–8. https://doi.org/10.1039/b918161j.

Vyas, M., B. Modhera, V. Vyas, and A. K. Sharma. 2011. "Performance Forecasting of Common Effluent Treatment Plant Parameters by Artificial Neural Network." *ARPN Journal of Engineering & Applied Sciences* 6, no. 1: 38–42.

Wang, J. L., and L. J. Xu. 2012. "Advanced Oxidation Processes for Wastewater Treatment: Formation of Hydroxyl Radical and Application." *Critical Reviews in Environmental Science & Technology* 42, no. 3: 251–325. https://doi.org/10.1080/10643389.2010.507698.

Wang, N., T. Zheng, G. Zhang, and P. Wang. 2016. "A Review on Fenton-Like Processes for Organic Wastewater Treatment." *Journal of Environmental Chemical Engineering* 4, no. 1: 762–87. https://doi.org/10.1016/j.jece.2015.12.016.

9 Policies and regulations

REGULATIONS AND POLICIES FOR THE RECYCLING AND REUSE OF WASTEWATER

At this crucial juncture, when the world is assessing the effects of various issues like climate change and how we can improve the water and sanitation systems to be made more sustainable through reducing the water-energy footprint and adapting the GHG mitigation strategies, there is need to adapt our approach to a "solutions-multiplying solutions" approach instead of a "problem-multiplying solutions" approach. And it is crucial for the planners and designers of water and wastewater recycling to incorporate the predicted effects of climate change and associated mitigating measures into overall policy and design. Due to their scale and untapped potential, water and wastewater recycling gives a huge opportunity for energy savings, but there are currently no regulations or incentives to limit energy use in the water and sanitation sector. The country needs a policy that includes energy efficiency in the water sector as a "necessity" rather than as an "option" or a "choice" at this point.

INTERNATIONAL

There are international frameworks and guidelines on recycling of wastewater and its further utilization, such as: (a) the international guideline of the World Health Organization (WHO), *Recommendations on Wastewater Treatment and Cultivation Limits of Wastewater Recycling in Agriculture and Aquaculture*; and (b) the guidelines and standards developed by California on water reuse, EPA (2012). They were the earliest to create recycling guidelines and standards to enhance the use of cleaned wastewater for specific uses.

WHO first established the guidelines for the reuse of wastewater, taking into account the methods of wastewater treatment and health protection measures (WHO, 1973) and primarily focused on describing appropriate levels of treatment required for various reuse applications. Through the use of available treatment methods and chlorination procedures, there is a possibility to achieve a bacteriological quality of 100 coliform organisms per 100 ml that is considered safe and as having limited health risk when used for the irrigation of food crops. These guidelines were further revised in 1989 according to more epidemiological and microbiological indicators of health risks usually associated with the use of untreated and treated wastewater (Table 9.1) (WHO, 1989). The WHO guidelines – further revised in 2006 – seems like a manual of good management practices to make certain that the use/reuse of wastewater in agriculture – mainly for irrigating crops, including food crops which are eaten raw – is safe and poses minimal health risks (WHO, 2006a, 2006b). To minimize the health risks from pathogens in wastewater when exposed to humans, the new the guidelines are developed that mainly focus on health-related goals instead of water

DOI: 10.1201/9781003407690-9

TABLE 9.1
The 1989 WHO guidelines for the use of treated wastewater in agriculture[a]

Category	Reuse conditions	Exposed group	Intestinal nematode[b] (arithmetic mean no. eggs per liter)[c]	Fecal coliforms (geometric mean no. per 100 ml)[c]	Wastewater treatment expected to achieve the required microbiological guideline
A	Irrigation of crops likely to be eaten uncooked, sports fields, public parks[d]	Workers, consumers, public	≤1	≤1000	A series of stabilization ponds are designed for the desired microbiological quality or equivalent treatment
B	Irrigation of cereal crops, industrial crops, fodder crops, pasture and trees[e]	Workers	≤1	No standard recommended	8–10 days retention in stabilization ponds or equivalent removal of helminth and fecal coliform
C	Localized irrigation of crops in category B if exposure to workers and the public does not occur	None	Not applicable	Not applicable	Pre-treatment as required by irrigation technology, but not less than primary sedimentation

[a] In some specific cases, the guidelines are modified and factors are considered like local epidemiological, sociocultural and environmental factors

[b] *Ascaris*, hookworms and *Trichuris* species

[c] Time of irrigation period

[d] A more stringent guideline is defined (≤ 200 fecal coliforms per 100 ml) for public lawns like hotel lawns, which are directly contacted by the public

[e] In the case of sprinkler irrigation, the irrigation is ceased before two weeks and the fruits are not picked from the ground

quality standards and provide different combinations of risk management options to meet these requirements.

Health-based goals in the 2006 WHO guidelines

The order of the approaches to protect human health in the 2006 guidelines is the following.

1. Establish the maximum additional burden of disease following the use of wastewater for irrigation of crops.
2. Determine the maximum number of pathogens that can be ingested without overcoming this acceptable disease burden.
3. Determine the number of pathogens which can be taken in different irrigation regimes for different types of crops through realistic human exposure scenarios.
4. Calculate the necessary pathogen reduction to be achieved, depending on the initial quality of the wastewater and the type of crop.
5. Select a combination of health-oriented controls to achieve needed reduction of pathogens.

Water reuse standards are created by U.S. Environmental Protection Agency (EPA) to supplement state, tribal, and other agency-developed regulations and policies in order to give national direction on water reuse regulations and program development. State and municipal governments in the United States are responsible for setting water reclamation and reuse requirements.

In 1980, the EPA Office of Research and Development commissioned a technical research report that included the first EPA recommendations for water reuse (EPA, 1980). It was updated in 1992 to help project planners and state regulators who were looking for advice from the EPA on water quality, use, and regulatory requirements for the development of treated water systems in various states (EPA, 1992). The update released in 2004 and 2012 primarily served the purpose of summarizing the water reuse guidelines with supporting data and information on utility benefit. The 2004 and 2012 EPA *Guidelines for Water Reuse* have an impact on worldwide regulatory bodies, particularly those in the United States. The new recommendations provide a national overview of reuse rules and explain some of the regional and state-specific peculiarities in the regulatory frameworks supporting the reuse. The EPA's water reuse guidelines have also had a significant global impact, as some nations even make reference to the text or take the guiding ideas of the 2004 guidelines as their own.

Reuse applications have grown significantly during the past ten years, reuse technology has advanced significantly and more countries have adopted reuse legislation or recommendations. Additionally, the increased demand for water supplies around the world has compelled planners to take non-traditional water sources into account while upholding environmental preservation. Through a cooperative research and development (CRADA) agreement with CDM and an inter-agency agreement with the U.S. Agency for International Development (USAID), the EPA created the 2012 *Guidelines for Water Reuse* to include these modifications and advancements in reuse.

NATIONAL

The policy makers and institutions of India comprises water recycling and reuse of wastewater in all the existing and upcoming programs related to water management, including the following.

1 The Planning Commission (as part of the water and waste management strategy in the 12th five-year plan).
2 The Ministry of Urban Development (as part of the National Urban Sanitation Policy [NUSP]) (http://moud.gov.in/NUSPpolicy), the National Mission on Sustainable Habitat (http://moud.gov.in/NMSH) and the Service Level Benchmarking (SLB) framework (http://moud.gov.in/servicelevel).
3 The Ministry of Water Resources (as part of the National Water Policy, 2012) (www.wrmin.nic.in/index1.asp?linkid=201&langid=1), the National Water Mission under the National Action Plan on Climate Change and the draft National Water Framework Law (www.wrmin.nic.in/index1.asp?linkid=220&langid=1).
4 The Ministry of Environment and Forests (as part of the National Environment Policy 2006 [http://envfor.nic.in/sites/default/files/introduction-nep2006e.pdf]).

The policies and guidelines in India place emphasis on the requirement for wastewater recycling, and various states and cities in India have such wastewater recycling projects, but they are mainly individually structured and designed in isolation at the regional level. To address the disparity in such projects, the Ministry of Urban Development (MoUD) recently developed specific and detailed guidelines based on management of wastewater recycling plants, which are given for treatment standards, design and operation. Besides this, there are plans to place tariffs on the sale of recycled wastewater for various other applications. Factors like urban sanitation, with wastewater recycling, are the focus of the ministry and it was included first time in the *Manual on Sewerage and Sewage Treatment* (CPHEEO, 2013).

CPHEEO, 2013

The Central Public Health and Environmental Engineering Organisation (CPHEEO) was set up in 1954 under the Ministry of Health, as per the recommendations of the Environmental Hygiene Committee that was responsible for all important sanitation programs for the country. Later, it was affiliated to MoUD in 1973–1974. The CPHEEO guides the union government, the states and the concerned ministries in the preparation of policies and providing technical expertise pertaining to new technologies, scheme evaluation or examination, and concerns with water supply and sanitation, such as municipal solid waste management, etc. Moreover, CPHEEO provides valuable notes on schemes by the agencies such as the World Bank/ Japan International Cooperation Agency (JICA)/Asian Development Bank (ADB)/ Kreditanstalt für Wiederaufbau (KfW)/Agence Française de Développement (AfD) or others in relation to above said issues.

CPHEEO earlier published the *Manual on Sewerage and Sewage Treatment* in 1993, which was further revised and updated in 2013. It was revised and updated according to new developments and changes related to technologies for sewage and sludge treatment, collection, transportation, and reuse for a variety of applications. The following objectives that have to be fulfilled, according to the revised and updated *Manual on Sewerage and Sewage Treatment.*

1 Better understanding of biological treatment mechanisms.
2 Use of advanced treatment methods in removal of specific constituents.
3 The increased focus on sludge management as a result of sewage treatment, as well as management of sewerage and sewage treatment in general.
4 Issuing permits with more stringent guidelines for the discharge and recycling of treated sewage.

Through providing guidance on how to choose options for selection of technologies for new infrastructure or upgrading current services, the existing manual (CPHEEO, 1993) was revised and updated (CPHEEO, 2013) to address some of those demands. It can be used for both small-scale localized initiatives and large-scale city-wide sanitation improvement programs. Following is a list of technologies with a variety of alternatives for offering techno-economic solutions while keeping in mind community health and environmental protection.

1 Decentralized sewerage networks
2 Treatment and management of sludge and septage
3 Advanced methods of sewerage and sewage treatment
4 Searching for new materials of pipe for sewerage construction
5 Manuals on recycling and reuse of treated wastewater
6 New standards for releasing treated sewage into reservoirs used for drinking

ESTABLISHING ENVIRONMENTAL POLLUTION STANDARDS AT THE STATE LEVEL

Before establishing urban cleanliness initiatives, the involved organizations should establish standards and implement them at the state level (in context with the general national standards), such as CPHEEO and BIS recommendations as listed in what follows.

1 *Environmental outcomes* such as State Pollution Control Board standards on effluent parameters, depleting water supplies, effects of climate change, use of on-site/decentralized sewage treatment technologies that run on less energy, distributed utilities, etc.
2 *Public health outcomes* such as state health departments.

3 *Processes* such as safe disposal of on-site septage, infrastructure (such as design standards), PHEDs/parastatals and coverage of the informal sector activities like sewage disposal, solid waste disposal, etc.
4 *Service delivery standards* such as by the urban development departments.
5 *Manpower issues* such as adequate pay, risk of hazards in the work and transparency of terms and conditions, as well as utilization of safer and more modern technology and provision of adequate safety equipment such as gloves, boots, masks, check-ups, regular health, medical and accident insurance, etc.
6 *The encouragement and awareness of citizens* should exist to adopt standards in accordance with public health and the environment. It is advised that states not only imitate but also establish standards beyond the national norms.

OVERVIEW OF CPHEEO MANUAL AND ITS PARTS

The latest CPHEEO *Manual* is divided into the following three interdependent parts.

1 *Part A: "Engineering"* includes the core methods and updated approaches in the direction of the shifting sanitation from on-site management to decentralized or conventional collection, transport, treatment and reuse of wastewater. Further, it is simplified to the level of the practicing engineer for daily supervision in the field in accepting the situation and coming out with a choice from a range of options to address. In order to improve the receiving environment, it also contains recent developments in sewage treatment and sludge and septage management. It is a straightforward instruction for the field engineer.
2 *Part B: "Operation and Maintenance"* explains the challenges with standardizing the financial and human resources. These are required to maintain the expensively constructed sewage and sanitation systems, and to prevent them from falling into disrepair for lack of formalized O&M criteria that would make it possible to handle related problems. These financial and related difficulties must be dealt with within the estimate stage itself in order to request a thorough approval of funding and staffing levels. To make the projects self-sustaining, this would also usher in the era of public–private partnerships. Along with occupational health risks and safety precautions for sanitation employees, this also addresses issues like standards for cleaning sewers and septic tanks. It is a straightforward guide for those looking for resources and those in charge of assigning them.
3 *Part C: "Management"* includes the latest methods of project delivery and project validation, which also provides a continuous model for the administration to use in predicting allocation deficiencies and implementing newer mechanisms. It is a tool for defending the project delivery method of choice and optimizing investments on need-based allocations as opposed to budget allocations that go unused and are turned over at the end of the fiscal year with no one using the money at that time.

It is a simple evolution of a humdrum strategy over the years. It is crucial to note at the outset of this description of Part A of the *Manual* that any trade names, technical nomenclatures, etc., that are quoted are merely included for the reader's convenience.

REFERENCES

CPHEEO. 1993. *Manual on Sewerage and Sewage Treatment*. New Delhi: Ministry of Urban Development.

CPHEEO. 2013. *Chapter 5. Manual on Sewerage and Sewage Treatment*. http://cpheeo.nic.in/WriteReadData/Cpheeo_Sewarage_Latest/PartAHighResolution/Chapter%205.pdf.

EPA, U. S. 1980. "Protocol Development: Criteria and Standards for Potable Reuse and Feasible." In *Alternatives*. Washington, DC: Environmental Protection Agency: 570/9–82–005.

EPA, U. S. 1992. *Guidelines for Water Reuse*. 625/R92004. Washington, DC: Environmental Protection Agency.

EPA, U. S. 2004. *Guidelines for Water Reuse*. 625/R-04/108. Washington, DC: Environmental Protection Agency.

EPA, U. S. 2012. *Guidelines for Water Reuse*. EPA/600/R-12/004. Washington, DC: USEPA Office of Wastewater Management.

WHO. 1973. *Reuse of Effluents: Methods of Wastewater Treatment and Health Safeguards and World Health Organization*. https://apps.who.int/iris/handle/10665/41032

WHO. 1989. *Health Guidelines for the Use of Wastewater in Agriculture and Aquaculture (Technical Report Series 778)*. Geneva: World Health Organization.

WHO. 2006a. "Guidelines for the Safe Use of Wastewater, Excreta and Greywater." In *Wastewater Use in Agriculture*, Vol. 2. Geneva: World Health Organization.

WHO. 2006b. "Guidelines for the Safe Use of Wastewater, Excreta and Greywater." In *Excreta and Greywater Use in Agriculture*, Vol. 4. Geneva: World Health Organization.

10 Opportunities

Solar powered wastewater recycling (SPWR) offers various economic and environmental benefits to municipalities and users (such as industrial and agricultural): (a) a safe and consistent water supply; (b) nutrient-rich wastewater; and (c) reduction in the costs of pumping groundwater. There is need to devise more sustainable business models for municipalities and consumers. These models should be categorized on the basis of users such as industry, agriculture and commercial institutions/entities who can work with municipalities to ensure financial viability, follow water allocation rules and support suburban agriculture.

Undoubtedly, SPWR is a win-win situation for all the stakeholders in water supply and wastewater treatment – i.e., government, municipalities, industries and consumers – but still, the following aspects need to be discussed for its effective and successful implementation.

WATER SUPPLY AND WASTEWATER TREATMENT

Climate change and its spatial and temporal consequences on water resource availability may aggravate worries about predicted gaps in water supply and demand. There are already issues with competing water needs and a shortage of freshwater in many Indian towns. Many of these cities are compelled to obtain their water from expensive or distant sources. Due to these difficulties, providing water in these cities for various uses is more expensive. Figure 10.1 demonstrates how in some Indian cities, it is becoming more expensive to supply water to businesses.

From evaluation of the state of wastewater generation and treatment in India's Class-I cities and Class-II towns (Over 70% of people live in these cities), it is estimated that about 38,255 MLD of wastewater was generated in 2009 alone (CPCB, 2009). In proportion to this, only about 11,788 MLD of wastewater treatment capacity has been implemented so far; that roughly accounts for 31% of total wastewater produced in these two groups of cities and towns. Further, the existing sewage treatment facilities in Class-I cities and towns are only running at roughly 72% of their planned capacity, resulting in the discharge of more than 75% of the sewage into different water and land areas without being treated, leading to significant environmental contamination and public health risks (CPCB, 2009). Due to this, 80% of the nation's surface water bodies are contaminated by the discharge of untreated or only partially treated wastewater onto land or surface waters, which is a major source of pollution.

The urban population is expected to increase by more than 50% from 377 million in 2011 to 590 million in 2030 at the present rates of population growth (1.7% per year) and urbanization (3% per decade); that leads to municipal wastewater will rise proportionately to around 60,000 MLD (MGI, 2010). It is estimated that by 2030, the entire amount of treated municipal wastewater (assuming that 80% of it can be

DOI: 10.1201/9781003407690-10

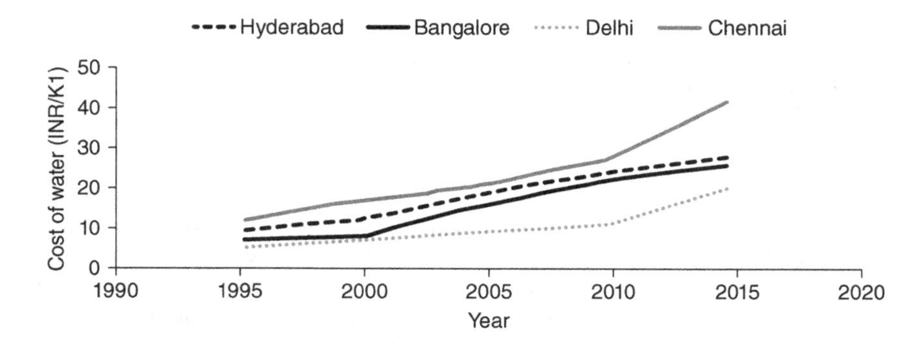

FIGURE 10.1 Cost of supplying water to industries in selected Indian cities

treated) will be available up to 17 BCM, which will increase the treated wastewater availability almost 400%. Upon proper collection, treatment, and recycling, this additional 17 BCM of treated wastewater resources will be sufficient to fulfill about 75% of the country's expected industrial needs in 2025 and nearly one-fifth of the world's projected drinking water demand.

TREATMENT TECHNOLOGY

Based on the physical boundary conditions and the intended use of the treated wastewater, techniques for the treatment and reuse of residential wastewater should be chosen. The demand for water of this quality will rise if wastewater is treated to a standard that is higher than what is necessary for it to be used safely in a required application. This will result in recovering the higher capital expenses, as well as increased operating and maintenance costs for service providers. Due to transportation cost savings and cutting-edge biological treatment methods, decentralized wastewater management systems typically have a lower resource footprint and consume less energy (30%–35% less energy on-site).

Based on studies related to various elements of wastewater treatment technologies, it is concluded that there is a need for different methods for the removal and treatment of contaminants. The traditional techniques for removing solid particles from wastewater include sedimentation, flocculation coagulation, flotation and filtration. For the removal of organic and inorganic components from biological treatment, advanced oxidation processes, adsorption, and membrane processes are preferable to conventional treatment. These modern water treatment techniques also have some drawbacks, such as those that it takes a lot of energy, induces fouling and produces a lot of byproducts. In addition to these, electrochemical techniques are gaining interest due to their high removal efficiencies, clean energy conversion, pollution avoidance due to the lack of emissions, and simple operation (Guo et al., 2022). Numerous electrochemical processes are explored, such as electro-oxidation, electro-flotation, electro-coagulation, electro-disinfection and electro-reduction (Koparal et al., 2002; Kumari et al., 2021; Guo et al., 2022).

SOLAR ENERGY

Solar energy may end up being the greatest option in the future for a variety of reasons: (a) the sun emits roughly 3.8×10^{-23} kW of solar energy each day, out of which the earth intercepts about 1.8×10^{-14} kW (Christopher, 2010); and (b) it is a very promising energy source due to consistent and increasing output efficiency over time compared to that of any alternative energy source. The earth receives solar energy in a variety of ways in addition to light and heat, the majority of which is absorbed, dispersed and reflected by clouds as it moves through the air. Being cost-free and abundant in nature, solar energy could meet the majority of the world's energy requirements.

Due to its ability to convert sunlight into electric currents in an eco-friendly manner, solar PV systems are gaining so much interest as alternative energy option (Makrides et al., 2010). Water pumps, lamps, battery chargers and delivery of electricity to utility grids are just a few applications for this source (Masoum et al., 2004; Meah et al., 2008; Bhandari et al., 2014; Enaganti et al., 2020). The production of PV modules has increased significantly at an incredible rate since the mid-1990s, suggesting the immense potential of PV systems for the present and the future (Parida et al., 2011). Further studies suggested that PV systems have not been affected by the weather in electricity generation. Ishii et al. (2013) and Pali and Vadhera (2020) found that PV systems can produce 80% of their potential energy capability in partly cloudy conditions, 50% in hazy or humid situations, and even produce 30% in really gloomy conditions.

In general, there are two types of PV systems: (a) standalone PV systems which need batteries to operate and which can be put in remote locations; and (b) grid-connected PV systems, which work with the local power grid to generate solar electricity that is then delivered by a separate energy supplier. The various financial incentives are provided for PV system installations in the public sector in an effort to encourage the adoption of renewable energy sources and raise environmental awareness. The components of a typical PV system are PV modules, batteries, a controller and an inverter. First, absorption of solar radiation takes place by a PV module, which transforms it into electric energy (direct current). From there, a regulator or controller transmits electric energy or power to the batteries and also protects the batteries from severe overcharging and discharge. During night and such times when there is little solar radiation, backup electricity is utilized through the stored solar energy. It is also observed that in the case of low solar radiation during the day, a PV setup without batteries is found to be suitable or purifying water. To save the cost of backup battery, it is advised that wastewater treatment can only be carried out during the day.

REUSE ISSUES

However, there are significant benefits of recycled wastewater: either it is reused in industries or reused in agriculture, but it is very different to recover the cost of operations and maintenance of a treatment plant in either situation. Through industrial reuse, there is sufficient revenue to cover the O&M expenses, while agricultural reuse provides negligible revenues for treatment plants. It is advised to encourage industrial reuse in all cities of a state, but it depends upon the availability of industrial entities around the particular cities.

In India, it is quite common practice to reuse untreated or partially treated wastewater in the agriculture, but wastewater recycling for industrial reuse is just emerging (Amerasinghe et al., 2013). To recover the expanses related to O&M of a treatment plant, there is need for regulators either to move from agriculture to industrial reuse or ban reuse in agriculture. It is very difficult to ban agriculture reuse due to the large number of dependent livelihoods. Further, if untreated or partially treated wastewater reused in agriculture is municipal wastewater and its treatment is inadequate, this would pose serious health consequences, especially diarrhea and helminth (parasitic worm) infections.

POLICY AND REGULATIONS

At present, water and wastewater facilities do not have policies or incentives to reduce energy consumption; thus, the country needs policies that incorporate energy efficiency in the water sector as a "necessity" rather than as an "option" or a "choice." There is a lack of a contemporary framework law for wastewater reutilization – i.e., CPCB standards in which the discharge of wastewater into surface water is covered by 33 criteria to be fulfilled, but wastewater irrigation through land application is only covered by eight criteria – but the fact is that by having these minimal criteria, wastewater is irrigated to the land unmonitored and unregulated, which is evidence of incompetence of state pollution control boards that are responsible for monitoring and controlling pollution in each state. This chapter outlines how different national policies and goals encourage the reuse of wastewater, but it is obvious that there are no defined criteria and regulations which encourage informal use of wastewater, which frequently has unfavorable effects.

Following are a few recommendations proposed by Hingorani (2011), some of which might be viewed as specifications for the reuse policy.

- A phased infrastructure plan that encourages the planning of an expansion of bulk water sources, along with the building of wastewater treatment facilities and newer sewer lines.
- Support decentralized treatment since sewerage networks are expensive to build.
- Reviewing the tariff structures that make secondary treated wastewater free to use and drinking water to be more expensive.
- The separation of "greywater" and "blackwater" is to be compulsory so that greywater can be utilized for gardening and toilet flushing before entering the sewers.

The Indian government recently introduced several initiatives to promote renewable energy and mitigate the climate change. *Smart City Mission* is one of these initiatives that attempt to ensure contemporary, effective urban infrastructure and efficient resource utilization in order to produce a clean, safe and sustainable environment. The *Atal Mission for Rehabilitation and Urban Transformation (AMRUT)* aims to improve fundamental infrastructure services such as the provision of drinking water and sewage and sewerage systems. In line for such programs and missions, the need for energy-efficient design gives improved cost effectiveness throughout the planning stage through investing in low-carbon water and wastewater management infrastructure.

The National Action Plan for Climate Change (NAPCC) also provides a list of existing and upcoming programs and strategies for reducing the effects of climate change and their adoption (MoST, 2008). NAPCC also intends to investigate how water, energy and climate change are related in keeping with its general objectives. The improvement of energy utilization in the urban water supply thus helps to meet cross-cutting objectives. There are a number of agendas emphasizing goals ranging from enhancing water delivery to efficient energy use.

CONCLUSION

Water supply and wastewater management challenges can be resolved in an economical and sustainable manner by treating and reusing wastewater. The price of sewage treatment involves many components such as the technology chosen, the possibility of economies of scale and the standard of the current treatment facilities. It is advised that financial incentives be offered to promote the reuse of secondary treated wastewater, including reasonable pricing of secondary treated water with alternative water supplies for the same or better quality, and by providing a continuous supply of water via secondary treated wastewater rather than an intermittent, rationed supply of freshwater.

Although there are a number of regulations for including wastewater reuse, their adherence is constrained by the absence of a legal framework. The wastewater reuse guidelines devised by many organizations should be merged and formulated as a legally enforceable national policy on reuse of wastewater to maximize the effectiveness of such a policy. The combination of GRIHA and IGBC principles offers a path forward for the successful integration of wastewater reuse programs. It can assist in integrating wastewater reuse from the project's inception and also monitor its execution to spot deviations or non-conformities. Adopting the guiding idea that "better grade water should not be used for applications that can tolerate lower quality" is urgently necessary for wastewater recycling and reuse (CPHEEO, 2013). The community's involvement and awareness campaigns can improve the acceptability of treated wastewater.

Julia Carney is quite lyrical in her description of the significance of protecting wastewater in protecting water: "Little drops of water, little grains of sand. Make the mighty ocean and the beautiful land."

REFERENCES

Amerasinghe, P., R. M. Bhardwaj, C. Scott, K. Jella, and F. Marshall. 2013. *Urban Wastewater and Agricultural Reuse Challenges in India*. 147. IWMI. Report No. 147. Colombo, Sri Lanka: International Water Management Institute.

Bhandari, B., K. T. Lee, C. S. Lee, C. K. Song, R. K. Maskey, and S. H. Ahn. 2014. "A Novel Off-Grid Hybrid Power System Comprised of Solar Photovoltaic, Wind, and Hydro Energy Sources." *Applied Energy* 133: 236–42. https://doi.org/10.1016/j.apenergy.2014.07.033.

Christopher J. Rhodes. 2010. "Solar Energy: Principles and Possibilities." *Science Progress* 93, no. 1: 37–112.

CPCB. 2009. *Status of Water Supply, Wastewater Generation and Treatment in Class-I Cities and Class-II Towns of India*. New Delhi: Central Pollution Control Board.

CPHEEO. 2013. *Chapter 7. Recycling and Reuse of Sewage* http://cpheeo.nic.in/WriteReadData/Cpheeo_Sewarage_Latest/PartAHighResolution/Chapter%207.pdf.

Enaganti, P. K., P. K. Dwivedi, A. K. Srivastava, and S. Goel. 2020. "Analysis of Submerged Amorphous, Mono-and Poly-crystalline Silicon Solar Cells Using Halogen Lamp and Comparison with Xenon Solar Simulator." *Solar Energy* 211: 744–52. https://doi.org/10.1016/j.solener.2020.10.025.

Guo, Zijing, Yang Zhang, Hui Jia, Jiaran Guo, Xia Meng, and Jie Wang. 2022. "Electrochemical Methods for Landfill Leachate Treatment: A Review on Electrocoagulation and Electrooxidation." *Science of the Total Environment* 806, no. 2: 150529. https://doi.org/10.1016/j.scitotenv.2021.150529.

Hingorani, P. 2011. The Economics of Municipal Sewage Water Recycling and Reuse in India. *India Infrastructure Report*, 312.

Ishii, T., K. Otani, T. Takashima, and Y. Xue. 2013. "Solar Spectral Influence on the Performance of Photovoltaic (PV) Modules Under Fine Weather and Cloudy Weather Conditions." *Progress in Photovoltaics: Research & Applications* 21, no. 4: 481–9.

Koparal, A. Savas, and Ulker Bakir Öğütveren. 2002. "Removal of Nitrate from Water by Electroreduction and Electrocoagulation." *Journal of Hazardous Materials* 89, no. 1: 83–94. https://doi.org/10.1016/s0304-3894(01)00301-6.

Kumari, Shweta, and R. Naresh Kumar. 2021. "River Water Treatment Using Electrocoagulation for Removal of Acetaminophen and Natural Organic Matter." *Chemosphere* 273: 128571. https://doi.org/10.1016/j.chemosphere.2020.128571.

Makrides, G., B. Zinsser, M. Norton, G. E. Georghiou, M. Schubert, and J. H. Werner. 2010. "Potential of Photovoltaic Systems in Countries with High Solar Irradiation." *Renewable & Sustainable Energy Reviews* 14, no. 2: 754–62. https://doi.org/10.1016/j.rser.2009.07.021.

Masoum, M. A. S., S. M. M. MousaviBadejani, and E. F. Fuchs. 2004. "Microprocessor-Controlled New Class of Optimal Battery Chargers for Photovoltaic Applications." *IEEE Transactions on Energy Conversion* 19, no. 3: 599–606. https://doi.org/10.1109/TEC.2004.827716.

Meah, K., S. Ula, and S. Barrett. 2008. "Solar Photovoltaic Water Pumping – Opportunities and Challenges." *Renewable & Sustainable Energy Reviews* 12, no. 4: 1162–75. https://doi.org/10.1016/j.rser.2006.10.020.

MGI. 2010. *India's Urban Awakening: Building Inclusive Cities Sustaining Economic Growth*. https://www.mckinsey.com/featured-insights/urbanization/urban-awakening-in-india

MoST. 2008. *National Action Plan on Climate Change, Ministry of Science and Technology, Government of India*. New Delhi: Government of India.

Pali, B. S., and S. Vadhera. 2020. "Uninterrupted Sustainable Power Generation at Constant Voltage Using Solar Photovoltaic with Pumped Storage." *Sustainable Energy Technologies & Assessments* 42: 100890. https://doi.org/10.1016/j.seta.2020.100890.

Parida, B., S. Iniyan, and R. Goic. 2011. "A Review of Solar Photovoltaic Technologies." *Renewable & Sustainable Energy Reviews* 15, no. 3: 1625–36. https://doi.org/10.1016/j.rser.2010.11.032.

Index

U

W

Y

Printed in the United States
by Baker & Taylor Publisher Services